ArcView® Developer's Guide

Amir H. Razavi

ArcView® Developer's Guide

Amir H. Razavi

Published by
OnWord Press
2530 Camino Entrada
Santa Fe, NM 87505-4835 USA

All rights reserved. No part of this book may be reproduced or transmitted in any form or by any means, electronic or mechanical, including photocopying, recording, or by any information storage and retrieval system without written permission from the publisher, except for the inclusion of brief quotations in a review.

Copyright © 1995 Amir H. Razavi
First Edition, 1995
SAN 694-0269

10 9 8 7 6 5 4 3 2

Printed in the United States of America

Library of Congress Cataloging-in-Publication Data

Razavi, Amir H., 1957-
   ArcView Developer's Guide / Amir Razavi.
      p.   cm.
   Includes index.
   ISBN 1-56690-059-X
   1. Computer software--Development. 2. ArcView. 3. Avenue
(Computer program language)   I. Title
QA76.76.D47R39   1995
910'.285'574--dc20   95-3828
                                      CIP

Trademarks

All terms mentioned in this book that are known to be trademarks or service marks have been appropriately capitalized. OnWord Press cannot attest to the accuracy of this information. Use of a term in this book should not be regarded as affecting the validity of any trademark or service mark.

ArcView and ARC/INFO are registered trademarks of the Environmental Systems Research Institute (ESRI), Inc., the world's leading supplier of Geographic Information Systems (GIS) software. Avenue is an ESRI trademark. OnWord Press is a registered trademark of High Mountain Press, Inc.

Warning and Disclaimer

This book is designed to provide information on customizing the ArcView program through Avenue, ESRI's object oriented programming language for ArcView. Every effort has been made to make the book as complete, accurate, and up to date as possible; however, no warranty or fitness is implied.

The information is provided on an "as-is" basis. The author and OnWord Press shall have neither liability nor responsibility to any person or entity with respect to any loss or damages in connection with or arising from the information contained in this book.

About the Author

Amir H. Razavi is a professional engineer registered in the states of Maryland and Virginia. He has a B.A. degree in civil engineering, and a Master's of information systems management from the George Washington University. Amir has been developing software since 1982, and has served as GIS manager for the Civil Rights Division at the Department of

iv   **ArcView Developer's Guide**

Justice. In 1994, he founded Razavi Application Developers, which specializes in ArcView and other applications. (The author can be reached on Compuserve at 71764,3262 and 71764.3262@compuserve.com on the Internet.)

Acknowledgments

This book would not have been a reality without the efforts and guidance of my editor, Barbara Kohl, and my project manager, David Talbott, at High Mountain Press. I am very thankful to both. I also appreciate ESRI's cooperation in providing the pre-release versions of ArcView 2.

Some of the geographic data files used in this book, including boundary files and major highways, were provided by Geographic Data Technology Inc., Lebanon, NH (800/331-7881). Other geographic files derive from the ArcView 2 software package. Matchware Technologies, Inc. (301/384-3997) professionals offered insight into the process of address matching. In order to provide real world examples in this book, I used portions of projects described in papers written by Dr. Robert C. Weih of Space Remote Sensing; Bennie Hutchins of Southwest Mississippi Resource Conservation & Development, Inc.; and Andrew Richardson of the California Department of Forestry. I am grateful to all.

I have been blessed with having a wonderful family that has supported me in every endeavor. Thus, I am dedicating this book to everyone in my family: to my wife, Tema, who inspired me to write this book; to my son, Dean, who cheered me along the way; to my parents, who have dedicated their lives to their children; and to my sisters, Homeria and Marva.

OnWord Press...

OnWord Press is dedicated to the fine art of professional documentation.

In addition to the author who developed the material for this book, other members of the OnWord Press team contributed their skills to make the book a reality. Thanks to the following people and other members of the OnWord Press team who contributed to the production and distribution of this book.

Dan Raker, President
Kate Hayward, Publisher
Gary Lange, Associate Publisher
David Talbott, Acquisitions Editor
Barbara Kohl, Project Editor
Laura Sanchez, Project Editor
Carol Leyba, Production Manager
Patrice Werner, Production Editor
Janet Leigh Dick, Marketing Director
Lynne Egensteiner, Cover designer, Illustrator
Robert Leyba, Production Assistant

CONTENTS

Introduction . . . . **xv**
    What is Avenue? xvi
    What is Object Oriented Programming? xvi
    Why Object Oriented Programming for Avenue? xvii
    How This Book Is Structured xviii
    How To Read this Book xviii
    Typographical Conventions xx

1 Structured Application Development 1
    Requirement Study 2
        The First Faire Bank's Requirements 3
    Prototyping 4
        Prototyping at the First Faire Bank 5
    Construction 6
        Construction of the First Faire Bank Application 7
    Structured Testing 7
        Structured Testing of the First Faire Bank Application . . . 8
    Conclusion 9

viii   ArcView Developer's Guide

2 Customizing the Interface 11

Elements of the Control Bar 12
    The Menu Bar 12
    The Pushbutton Bar 12
    The Tool Bar 13
Creating A Menu System 13
    Adding and Organizing Menu Options and Items 14
    Executing a Task Through a Menu 16
    Disabling and Hiding a Menu Item 17
    Assigning Access and Shortcut Keys 19
Creating Pushbuttons 20
    Adding and Organizing Pushbuttons 21
    Executing a Task Through a Pushbutton 23
    Disabling and Hiding Pushbuttons 24
    Adding a Help Line to Pushbuttons 25
Creating Tool Buttons 26
    Adding and Organizing Tool Buttons 27
    Executing a Task Sequence Through a Tool Button . . . 29
    Associating a Cursor Shape to a Tool Button 30
    Disabling and Hiding a Tool Button 31
    Adding a Help Line to Tool Buttons 32
Startup and Shutdown Scripts 33
Saving the Customized Interface 34
More To Come 35

3 Avenue Building Blocks 37

Avenue Characteristics 37
    Encapsulation 38
    Polymorphism 38
    Inheritance 38
Basic Building Blocks 39
    Classes 39
    Objects 39
    Requests 40

Contents **ix**

Interaction Between the Building Blocks 40

    Creating New Objects 40

    Making a Request 41

Creating a Script 44

    Using the ArcView Script Editor 44

    Using a System Editor 45

    Where to Start 46

    Accessing System Scripts 47

Script Testing. 47

    Script Compiling 48

    Running Your Script. 49

    Fixing Run-Time Errors 49

4 Avenue Programming Language 51

Programming Elements 51

    Referencing Objects with Variables. 52

    Using Literals. 53

    Using Lists and Dictionaries 54

    Controlling Program Flow 56

    Documenting with Comment Lines. 61

Event Programming 61

    Checking for a Condition 62

    Checking for Active Objects 63

Interaction Between Programs 64

    Calling Other Programs 64

    Passing Objects Between Programs 66

Accessing Files. 67

    The File Name Object 68

    Creating, Opening, and Closing a File 69

    Reading and Writing a File. 71

User Dialog 72

    Displaying Information, Warnings, and Errors. 72

    Getting an Input 73

    Getting a File Name. 75

Upcoming Events. 77

x   ArcView Developer's Guide

5 Programming the Project Window 79

Setting Up an Application 80
The Application Startup Script 82
Setting Project Parameters 83
Finding a Document 84
Arranging Document Windows and Icons 84
Displaying a Help Message 85
Displaying the Status Bar 86
Tutorial: Creating an Application 88
Development Overview 88
Stage 1: The Application Documents 89
Stage 2: Customize the Interface 90
Stage 3: Avenue Scripts 95
Testing Your Application 102
Your Learning Curve 102

6 Programming View Documents 105

Application Overview 105
The View Document 107
Creating a New View 108
Displaying a View Document 109
Setting View Properties 109
Setting the Display Extent 110
Graphical Elements 110
The Graphics List 115
Drawing Graphical Elements 115

7 Programming Themes 119

Adding and Displaying Themes 119
Data Source 122
Legends 124
Copying 125
Using Queries 126
Selecting Features 127

Point Selection. 128
Line Selection 129
Box Selection. 129
Polygon Selection 130

8 Programming Table Documents. 131

Assigning Polygons to Territories 131
Table Document 137
Creating a Table. 138
Displaying a Table. 138
Setting Table Properties 140
Joining Tables 141
Sorting Records 142
Printing Tables. 142

9 Accessing Databases 145

Creating File Based Database Objects. 146
Reading Records and Fields 150
Accessing an SQL Database 153
Using ArcStorm 155

10 Programming Chart Documents. 159

Programming Chart Documents 159
Chart Document 163
Creating a Chart. 163
Setting Chart Type and Style 165
Setting Chart Properties 167
Working with Data Elements 170
Accessing Records 170
Accessing Fields. 172
Identifying a Record. 172
Finding a Record by Matching a String 172
Printing Charts. 173

xii   ArcView Developer's Guide

11 Programming Layout Documents 175

LayoutDocument Script 176
Layout Documents. 183
   Creating a Layout. 184
   Layout Display. 184
   Setting Layout Properties 185
   Using the Graphic List 187
Framing a View Document. 190
   Adding a Scale Bar 192
   Adding Legends. 193
   Adding a North Arrow 193
Framing a Table or Chart Document 194
Framing a Picture 195
Printing the Layout 198

12 Application Installation 199

Protecting Your Scripts. 200
Distributing Objects. 202
Single User Installation 203
Network Installation 205

13 Address Matching 207

Geocoding with ArcView. 208
   Making a Theme Matchable. 208
   Adding Address Events 210
   Creating an Events Theme 210
   Processing Unmatched Events 210
Geocoding with Avenue 211
Making a Theme Matchable. 216
   Selecting a Theme 217
   Selecting an Address Style 217
   Associating Attribute Fields 219
The Geographic Dimension of Data 222
   Creating a MatchSource Object. 225
Creating an Address Events Theme 226
   Retrieving the Match Source. 226

Selecting the Event Table 227
Initializing a Geocode Theme. 228
Matching Addresses 228
Creating the Address Event Theme 232

14 Integration 233

Accessing the Clipboard. 234
Accessing the Operating System. 236
Accessing Environment Variables 236
Issuing Operating System Commands. 237
Executing AMLs 237
Implementing Dynamic Data Exchange 238
ArcView as DDE Client 239
ArcView as DDE Server 240
Implementing Remote Procedure Call. 240
ArcView as RPC Client 241
ArcView as RPC Server. 243

Appendix A   Avenue Class Hierarchy 245

Appendix B   Avenue Reserved Words. 255

Index 259

Introduction

ArcView, by the Environmental Systems Research Institute (ESRI), is a new approach to spatial analysis. Prior to the advent of ArcView, exploring spatial data involved extensive training and often extensive programming. Consequently, technicians usually performed analysis or developed customized data query and analysis routines for business users. By providing an intuitive graphical user interface, ArcView goes a long way toward bringing non-technical users closer to Geographic Information Systems (GIS).

Similar to many other commercial software packages, ArcView is designed to appeal to a broad range of users. As a result, sooner or later you will reach a point where customization of ArcView is desirable to fit your unique needs. By using Avenue, ArcView's programming language, you can customize the program and further extend its power.

What is Avenue?

Avenue is an object oriented programming (OOP) language, the most recent advance in software application development technology. With Avenue you can create a new interface for ArcView or customize the current one, automate repetitive tasks, and write complete query and analysis applications. Avenue is equipped with a library of classes that represent the objects found in ArcView. The program executes tasks by accessing and manipulating these objects.

What is Object Oriented Programming?

A programming paradigm is the basic approach to writing a program, and different paradigms satisfy different types of software applications. The procedural paradigm is well known. Other paradigm types include objected oriented, rule based, and visual.

One way to describe object oriented programming is through metaphor. Computer science is replete with metaphors, such as "memory" and "windows". Although what metaphors represent in computer science is far from what they resemble in the physical world, they are useful for understanding software functionality. The basic metaphors for object oriented programming are "objects", "classes", and "requests".

An object consists of both code and data, and can be conceptualized as "smart code" or "active data". A class is the blueprint to create a functioning object, but is also an object with very limited capabilities. An analogy here would be that while a class provides the genetic code for creating hens, the actual hen object, not its class, lays eggs. Next, objects communicate, such as in an object receiving a request, or objects exchanging requests with each other. Requests ask objects to perform tasks, and tasks are the functionalities embedded in the object.

Object oriented languages are categorized as "pure" or "hybrid" systems. In the case of C++, a hybrid object oriented language, objects coexist within a procedural programming language. In pure object oriented languages, such as Smalltalk and Avenue, everything is an object. While Avenue does not implement all object oriented programming techniques, the current version is impressively powerful and can satisfy nearly all analysis applications.

Why Object Oriented Programming for Avenue?

The object oriented paradigm has addressed many problems inherent in the procedural approach to programming. The principal advantage of the object oriented language is its ability to handle complexity in a transparent manner. Under the object oriented paradigm, functionalities are placed inside the objects and out of the programmer's sight. Consequently, to execute a specific functionality, the programmer simply makes a request to the object.

When you become familiarized with Avenue and the object oriented paradigm, you may notice that this paradigm eliminates redundant code. You may also notice that you save time by building a program with objects, and that the quality of your programs has improved since object functions are protected from any code you write. Finally, you may actually become convinced that the object oriented approach is a superior way to program.

Because Avenue is an optional product and is separate from the ArcView package, you must have an Avenue license in order to apply anything in this book. To verify whether Avenue is on your system, start ArcView and examine the Project window. ArcView components are listed in this window starting with Views. If the Scripts component is present, Avenue has been installed. Installation is sequenced in that ArcView precedes Avenue. When installing the software from a CD-ROM, Avenue installation must be specifically requested.

How This Book Is Structured

Chapters 1 through 4 introduce the concepts of application development, ArcView interface customization, and Avenue as object oriented programming language. Chapters 5 through 11 are dedicated to how the diverse parts of ArcView are accessed and programmed using Avenue. Chapters 12 through 14 focus on advanced topics and housekeeping.

There are two appendices. Appendix A contains the Avenue class hierarchy. Appendix B provides a list of words and expressions reserved for use by the Avenue program.

Finally, there is a complete index at the end of the book.

How To Read this Book

Most people involved in developing GIS applications are professional geographers, cartographers, surveyors, planners, geologists, etc. If you are among the ranks of these professionals, then you know that there is barely enough time to keep up with the technological advances in your own field, much less software engineering. However, if you have written programs in procedural languages such as Basic, COBOL, C, or AML (ESRI's Arc Macro Language, used in ARC/INFO), you will know that object oriented languages constitute a new and different approach to programming.

This book attempts to accelerate the process of learning Avenue. After describing the basics of object oriented programming, the book then demonstrates how object oriented programming is implemented in ArcView and how you can take advantage of it.

You do not have to be a seasoned programmer to understand the *ArcView Developer's Guide*. However, this book is about programming with Avenue, and is not a programming tutorial. Next, you do not have to be a GIS expert to benefit from the *Guide*, but effective application developers understand the target environment. If you plan to extend the power of ArcView beyond what you are able to do with the package as shipped from ESRI, this book is for you.

Because this book is about application development, a discussion of the application development process could not be avoided. Chapter 1 reviews the structured approach to application development. If you are a seasoned software engineer, this chapter can serve you as a quick review.

Everyone else should read Chapter 1 in order to recognize that there is a method to application development. Countless volumes about development approaches are available, and the few pages appearing here barely scratch the surface. The purpose of this chapter is to whet your appetite for additional readings on the subject.

Avenue scripts are implemented through customized controls such as menu items and pushbuttons. Chapter 2, "Customizing the Interface," discusses these controls and explains how they are merged with Avenue scripts. For both experienced and fledgling programmers, this chapter provides a preview of the task environment that you wish to modify through Avenue.

If you are new to object oriented programming languages, read Chapter 3, "Avenue Building Blocks," very carefully. The chapter describes the basics of object oriented programming and how they are implemented in Avenue.

Chapter 4, "Avenue Programming Language," presents Avenue's programming elements. You must read this chapter if you are new to programming. Experienced programmers should review this chapter to become familiar with the Avenue language.

Chapters 5 through 11 demonstrate how various elements of ArcView are accessed and programmed using Avenue. The information in these chapters will be applied in most of your application programs.

If you plan to distribute your customized ArcView application, you should study Chapter 12. This chapter discusses how to protect your application when installed at user sites.

Chapters 13 and 14 are advanced topics that you may want to read after you have gained some experience in programming with Avenue. Address matching with Avenue is explained in Chapter 13, and the last chapter focuses on integration of ArcView into other applications.

xx   Introduction

Typographical Conventions

✔ **NOTE:** Information on Avenue features and procedures that may not straightforward or intuitive may appear in a note.

✘ **TIP:** Tips are the fruit of experience, and are aimed at saving you time and stress.

Text (code) that you type into an Avenue script is shown in a monospaced typeface:

```
if (nil = theView) then
```

❏ Menu items, tool buttons, and other functions in ArcView are capitalized and shown in boldface:

The **Category** option must be set to **Menu**.

❏ Avenue programming elements, such as requests and object classes are shown in italicized boldface:

The ***Make*** request to ***TextFile*** or ***LineFile*** classes can create new files or open existing files.

❏ Names for files, directories, variables, and objects, and user-derived input of diverse kinds appear in italics:

Avenue scripts in this chapter use the *USA* coverage bundled with the ArcView software. The coverage is located in ArcView's *AVDATA* directory.

Structured Application Development

Similar to computer applications in other fields, GIS applications must be developed according to a structured methodology. If you plan to write short Avenue scripts to automate a few tasks, development methodology may not be for you. If you plan to develop complete applications or implement new systems, following a structured methodology dramatically increases the probability of a successful product.

For many people "methodology" implies extensive overhead, extraneous tasks, and the stifling of creativity. In reality, however, adopting a methodology simply means using certain techniques and executing certain tasks that will produce the best results.

You may be familiar with one of several development methodologies tested over the past several years, particularly if you have developed ARC/INFO applications. The methodology discussed below is both a combination of and variation from methods used in developing GUIs (Graphical User Interfaces) and client/server application software. It is

2   Structured Application Development

also somewhat different from traditional methods used in developing ARC/INFO applications because ArcView is a new generation product.

The proposed structured methodology for developing ArcView applications is comprised of the following stages:

❏ Requirement study

❏ Prototyping

❏ Construction

❏ Structured testing

In the remainder of this chapter, the four stages are described and illustrated through a hypothetical scenario. The scenario assumes that we are GIS consultants to a large metropolitan bank for developing an application showing characteristics of mortgage applicants and recipients within designated geographic areas.

Because the techniques and procedures for each stage in the recommended methodology have been applied in developing many software applications, detailed information is available in numerous publications on software development methodologies. Next, it should be mentioned that development methodology is only one part of implementing a GIS project. Other issues such as hardware, data sets, staffing and procedures must also be considered, but are beyond the scope of this book.

Requirement Study

The requirement study stage begins upon recognizing that a solution is required for a particular problem. The purpose of a requirement study is to develop a specification document describing what the software is to accomplish, but without explaining how the software will work.

The specification document is the bedrock of your application. The more accurate and complete the foundation, the better your final product will be. Managers and analysts often ignore this stage in favor of writing program code because they prefer seeing a tangible outcome as early as possible. Nevertheless, experienced software developers would assert that you are doomed to fail if you ignore a study of software requirements.

Two distinct activities occur during the requirement study phase: problem analysis and product description. These activities are not carried out in serial fashion, but rather product description evolves as problem analysis progresses.

Because the purpose of analysis is to acquire a complete understanding of the problem, most of the analyst's time is spent in meetings with people who are knowledgeable about the problem. In the course of product description, the analyst develops the specification document that explains exactly what a software application can do to resolve the now understood problem.

The First Faire Bank's Requirements

During a meeting with First Faire Bank (FFB) managers, we learn that the bank president Donald Zinger is concerned about recent media attention on alleged discriminatory lending practices by other banks in surrounding counties. Since FFB management is genuinely committed to a nondiscriminatory lending policy, Zinger wants to ensure that no one will be able to identify patterns which even remotely suggest discrimination. The president is aware that GIS can be used for monitoring any number of variables within specific geographic areas.

The requirement phase begins by identifying the type of evidence an adversary could use when claiming that FFB engages in discriminatory lending practices. FFB's chief legal counsel Mary Lawsome explains that in some cases, branch banks allegedly approved disproportionately fewer mortgage contracts in areas of high minority population density than in low density areas. In these instances, average household income was similar in all locales studied.

Next, we learn that annual reports on mortgage applicants generated by FFB include ethnicity and a variety of socio-economic status variables. Applicants are also geographically referenced by census tract. In addition, the bank produces quarterly statistical reports on the distribution of active mortgages by ethnic group. However, because most bank managers consider these reports to be complicated, they do not bother to study them.

At this juncture, we decide that FFB needs a GIS application which presents branch offices, mortgage counts, and mortgage values over a map displaying minority population density. Required data components

4 Structured Application Development

include census tract boundaries, demographic data, branch locations, and mortgages (applicants, approved contracts, and dollar amounts). Equipped with these data sets, the application could produce demographic maps, branch location maps, pie charts of mortgage numbers and amounts by census tract, or any combination of the above. We then document the requirement study and distribute copies for review to the president, general manager, and chief counsel.

Prototyping

Under traditional methodologies, analysts would write the design specification document following the requirement study phase. Next, users were requested to approve the specification document before analysts commenced producing program code. However, because users could rarely fully comprehend the paper specifications, user approval at this phase in the process was generally meaningless.

The purpose of prototyping is to quickly develop a rough version of the desired system. Users can more easily comprehend the application by reviewing the prototype than by studying paper specifications. A prototype can also be adjusted and modified to show various design alternatives.

In ArcView, prototyping begins with customization of the interface controls and project components. Try to avoid writing Avenue scripts for the application logic as part of the prototyping. Instead, identify the required scripts and write a short description for each, and define and diagram the data flow. Review your prototype with users and describe how the application will behave once the scripts are developed and incorporated. Repeat this process until users are satisfied with your description of the desired system.

As you become more experienced with ArcView, Avenue, and this methodology, user participation in developing the prototype can accelerate the process. In fact, you could combine the requirement phase with prototyping for simple applications if you have strong user participation.

Prototyping at the First Faire Bank

Within a few days of submitting the requirement document, FFB managers contacted us with their responses. Fortunately, everyone likes the proposed GIS application. Because the legal office is to be the primary user, chief counsel Lawsome assigns a member of her staff, Richard Atty, to work with us in developing the application. We also request that the general manager assign a financial analyst to assist us in deciphering the annual mortgage reports.

A quick survey of the bank's legal office staff reveals that while everyone uses personal computers and GUI applications, no one is familiar with the GIS concept. Consequently, we present a half-day seminar to the legal office staff consisting of an introduction to GIS, and a demonstration of ArcView 2. As a courtesy, we invite the bank president and general manager, who accept and attend the seminar in the company of several other management staff members.

After the seminar, we spend a few hours with Atty in order to demonstrate selected ArcView 2 applications in detail. Meanwhile, we explain how the software can be customized to suit the bank's needs.

We then decide to begin prototype design by customizing the interface and preparing sample data. After the first cut of the prototype is complete, we rehearse a scenario to take users through the application. Next, we meet with Atty to review the prototype. He is very excited about our work, but makes no critical comments or suggestions. Either we have done a terrific job, or the user is not prepared to participate. We decide to be cautious, and give Atty a few extra days to digest the prototype. We make it easy for him to access the prototype, and we remain available to answer questions by phone.

Meanwhile, we stay in touch with Atty to make sure he is experimenting with the prototype and has not forgotten us. Within several days, Atty breaks his code of silence and starts asking questions. Questions eventually lead to suggestions and recommendations. As a result, we prepare a second version of the prototype and demonstrate it for all legal office staff. The demonstration generates a few comments on required and desirable functionalities. After we have incorporated the results of all commentary into the third version of the prototype, the legal office staff agrees to accept the prototype as is. At this point, we

6   Structured Application Development

print the customized interface and document the functionality that each interface control must provide.

Construction

In the construction phase, the Avenue scripts required for the application are developed. Before writing code, standards for current and future applications should be established. The extent and range of the standards depend on your application. Script documentation and naming conventions are minimum standards. You should also determine how common tasks such as error handling and data sharing procedures are dealt with.

Similar to many GUI applications, ArcView applications consist of numerous scripts that have no apparent relation to each other. The scripts are usually linked to a control interface and perform their work independently. Before writing each script, document in plain English how the tasks are carried out. This practice will help you develop clean and efficient scripts. Keep in mind that because Avenue is not a procedural language, ArcView does not respond to requests in a serial fashion.

Applications written with procedural programming languages control user access to features.or functions. In contrast, users control access to features or functions in ArcView applications written with Avenue, and in other GUI applications.

For example, a GIS application could show sewer pipes installed before a certain year within a specified zip code area. In ARC/INFO, such an application is written with AML, a procedural language. The ARC/INFO application asks for the zip code before zooming into the area. The user is then asked to input the year before the application displays the pipes.

The same application in ArcView would present menu items and pushbuttons to perform each task independently of other tasks. In brief, the user controls the order of the process. For instance, the user could choose to display all the pipes first, and then provide a year to select from the displayed pipes. At this point, the user can choose to zoom into a particular zip code area.

In AML, tasks are carried out serially, which means that one task must be performed before the next in the series can begin. In Avenue, tasks are carried out by sending a request to an object. (Requests and objects are discussed in detail in Chapter 3.) The important concept here is that the object receiving the request will carry out the task. Depending on the request, Avenue may not wait for the object to finish its task before moving on to the next task.

For example, when an Avenue script designed to change the map scale iterates through a loop five times, the map scale is changed every time. However, the map display you see is the result of only the final scale change. Every time the scale is changed, a request is sent to update the map with the new scale. Avenue does not wait for the map on the screen to be updated; the scale is changed and another request is sent that overwrites the previous one.

Construction of the First Faire Bank Application

While waiting on user reviews during the prototype phase, we prepared for the construction phase. Because the bank has no previous experience with GIS applications, specific GIS-related standards do not exist. However, FFB programmers have established GUI standards for their Windows applications. By the close of the prototyping phase, we have reviewed the bank's GUI standards along with the file structure of the required data as it is maintained on the bank's mainframe. We have also added our standards for error handling and naming conventions.

We are now ready to create Avenue scripts. The document we generated at the end of the prototyping phase can organize our progress through the construction phase. We will check off each functionality as we complete respective Avenue scripts. Involvement with the users at this stage is limited to clarifying or resolving incidental issues.

Structured Testing

The purpose of testing is to identify application software defects. Resolving as many defects as possible before delivery is important because people can lose faith in your ability as a developer if your

8   Structured Application Development

software product is loaded with defects. Therefore, testing must be viewed as a destructive process in the search for software errors.

There are three levels of software testing. "Unit testing" is executed by the programmer to ensure that each Avenue script behaves as expected. "System testing" is also carried out by the programmer except in cases involving large projects where an independent test team accepts the responsibility. The purpose of system testing is to determine whether various scripts work together properly and whether the system meets requirement specifications. "User acceptance testing" is executed by users to determine if the application meets their needs.

Structured Testing of the First Faire Bank Application

Because we are seasoned developers we do not succumb to unstructured testing, a common trap for many novice programmers. Unstructured testing produces favorable results because programmers unconsciously test in a manner that avoids software failure.

Instead, we begin by developing a plan for testing each script and the entire application for boundary data, typical data, and complete code coverage. Boundary data refers to cases in which data values extend beyond what is considered normal or acceptable. Complete code coverage refers to executing the script under various conditions to ensure that each line of script is executed at least once.

Subsequently, we develop test scripts for each unit and the entire application. Each script describes the input, expected output and procedures. Next, each script is executed, and the output is compared with the expected output. We carry out each test script, recording the output and comparing it with the expected output.

A few inconsistencies are investigated and corrected. Upon submitting the application to the FFB legal office staff for testing, we are informed a week later that it will be officially drafted into use as soon as we change three minor interface features.

Conclusion

The First Faire Bank scenario was perhaps a bit smoother than most structured application developments in the real world. However, the stages described here constitute the bedrock of effective application development, and cannot be overemphasized.

Chapter 2

Customizing the Interface

During the process of application development, you should consider human interaction with your program, or how users control your application's process through the program interface. Interaction complexity is directly related to the flexibility of your application. If the user is required only to start the program and provide a few parameters, the interface can be simple. However, if the user needs interactive tools to analyze and direct the application, the interface will be more complex.

Next, users judge a program's friendliness by its interface. If the interface does not provide feedback, or is not intuitive, users may perceive the program as unfriendly. On the other hand, a well-designed interface can compensate for other application shortcomings such as slow response time.

ArcView gives you the ability to create and customize your own interface. Therefore, it is up to you to design the best interface possible for your application. In this chapter we discuss the steps in creating an interface tailored to your application.

Elements of the Control Bar

Interaction with ArcView takes place through the control bar. The control bar consists of three lines containing the menu, pushbuttons, and tool buttons, respectively. Each ArcView document has a control bar, which becomes available when the document is active. The following illustration shows the control bar for **View** documents.

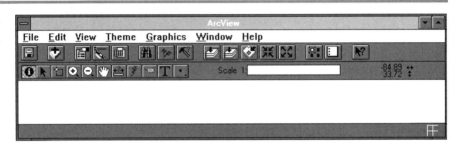

The control bar for View documents.

The Menu Bar

The line of menu options is usually displayed at the top of the control bar. Additional menu items for each option are revealed by clicking the pointer on the option. Pull-down menu items offer the user an easy method of interaction. A well-designed set of menu bars makes the application more consistent and easier to learn.

Menu names must be meaningful and indicative of their actions. You should establish and follow a set of standards for naming and grouping items. Keep in mind that users of various systems are accustomed to certain naming and grouping conventions. For instance, users of Microsoft Windows generally expect **File** to be the first menu option, with pull-down items such as **New**, **Open**, **Print**, and **Exit**.

The Pushbutton Bar

Pushbuttons create a faster method for executing the same commands available through the menu items. Generally, pushbutton bars consist of frequently used menu items. Novice users often prefer menu options, while experienced users like pushbuttons.

The Tool Bar

The tool bar is a set of mutually exclusive tool buttons that associate a specific procedure with the cursor. Once selected, a tool button remains depressed until deselected. Because tool buttons are mutually exclusive, only one tool button can be selected at a time. Thus, selection of a new button deselects the current one. When a user selects a tool button, the cursor shape changes, preferably to the same icon that marks the button. A cursor shape change indicates that a tool button is active.

Creating A Menu System

When creating a menu system, you will be working extensively with the **Customize** dialog box. To access the Customize dialog box, make the **Project** window active by clicking the pointer on its title bar. Select the Project menu option from the control bar to display a pull-down menu, and then select the Customize menu item from the pull-down menu. As shown in the following illustration, selecting the Customize menu item will display the Customize dialog box.

14   Customizing the Interface

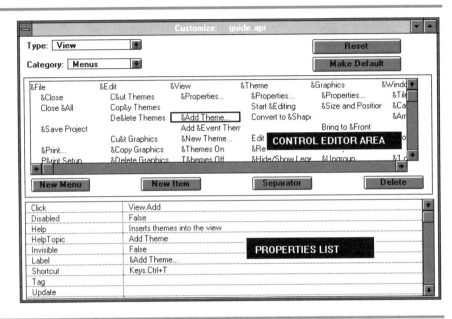

The Customize dialog box for a view document.

Each ArcView document type has its own menu bar. Select the menu bar that you want to customize by choosing the ArcView document from the **Type** option at the top of the **Customize** dialog box. The **Category** option immediately below must be set to **Menu**. The current menu specification is displayed in the **Control Editor** area, with menu options in the top row. The menu items pertaining to each option are indented and appear below the option name.

Adding and Organizing Menu Options and Items

To select a menu option from the top row of the **Control Editor** area, simply click on it. A frame appears around the name to identify the selected option. Create a new menu by clicking on the **New Menu** button. Any new menu will be created to the right of the selected one. To place a menu option at the beginning of the menu bar, create it, then move it to the beginning. You can move a menu option and associated items by dragging the option to its new location. If you drag the option and drop it on top of an existing one, ArcView places the new option to the left

Creating A Menu System   15

of the existing one. You can expand the Control Editor area by maximizing the **Customize** dialog box.

There is no limit on the number of menu options; ArcView creates a second line of the menu bar if all the options do not fit on one line. However, too many menu options could reduce readability and intuitiveness.

✗   TIP: Limit the number of your menu options to seven, and never create more than 12 for a single menu.

Menu items are manipulated in much the same way as menu options. Create a new menu item by first selecting a menu option and then clicking the pointer on the **New Item** button. The new item will appear as the last entry on the option's list of items. If you select an existing menu item before creating a new one, the new menu item will appear after the existing item. You can move the items around within an option by dragging and dropping, but you cannot drag an item out of its option.

You might consider using a separator between groups of items to improve readability. To place a separator after an item, select that item and click the **Separator** button.

Because ArcView customization does not allow more than one tier of menus, menu items cannot be used to open a second list of menu items. Keep this limitation in mind when designing your menu system. If you absolutely must have additional menu tiers, you can use a **Choice Message Box** to emulate the second tier. However, the look and feel of Message Boxes are not consistent with menu systems, and this inconsistency is undesirable in an application. Message Boxes are discussed in Chapter 3.

ArcView initially assigns **Menu** and **Item** as the names of created options and items. The defaults should be changed to more meaningful names related to the purpose or action of the menu. To change a name, double-click on the **Label** line inside the **Properties** list. A Label dialog box, shown in the following illustration, appears to accept the new name. There is no limit on the number of characters for the name, and all printable characters, including spaces, are acceptable. Lengthy names, however, make the application unfriendly.

The Label dialog box.

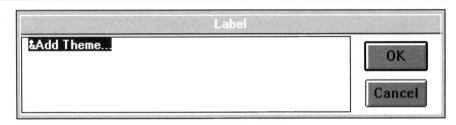

✘ **TIP:** Use a singular noun, such as **File** or **View**, to name menu options. Make all item options either a verb or a noun, such as **Copy** or **Tile**. Add an ellipsis (...) to the name of a menu item if its selection results in a dialog box.

Executing a Task Through a Menu

A task is executed through a menu item by associating a script to the item. You can use a system script that is already available or you can write your own. Chapters 3 and 4 explain how to write new scripts.

First, select the desired menu item. Inside the **Properties** list of the **Customize** dialog box, double-click on the **Click** property line to display the **Script Manager** dialog box. In this dialog box, shown in the following figure, available scripts are listed. The list includes both system and programmer-created scripts. Select the desired script from the list and click the pointer on the OK button. The name of the selected script appears in the Properties list of the menu item. You can review the contents of system scripts or create new ones based on existing scripts. This process is discussed in Chapter 3.

The Script Manager dialog box.

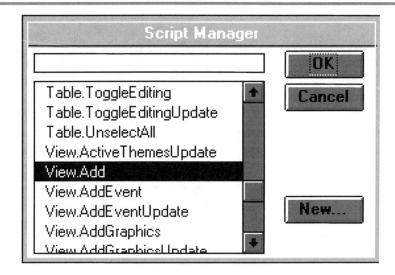

Disabling and Hiding a Menu Item

Under certain circumstances your application may require that specific menu options or items be unavailable for selection. In this case, you can either disable the item or make it invisible. A disabled item is displayed in gray but cannot be selected, while an invisible item is not displayed. Menu options or items are automatically rearranged to fill in the gap resulting from an invisible menu. The following figure shows the display change when **Delete** is disabled, and **Window** and **Import...** are made invisible.

18   Customizing the Interface

Disabling and making invisible menu options and items.

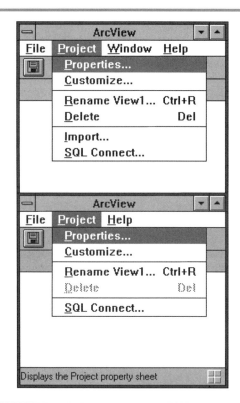

You can disable or make invisible both menu options and menu items. When an option is affected, its items become inaccessible. Select the option or item you wish to disable or make invisible from the Control Editor area of the Customize dialog box. The Disabled and Invisible property lines in the Property list indicate the status of the menu option or item. A False value for the Disabled property means that the menu option or item is not disabled and can be selected. A False value for the Invisible property means that the menu option or item is displayed.

The value of a **Disabled** or **Invisible** property can be changed by double-clicking its property line, which toggles between True and False. The "Event Programming" section in Chapter 4 explains how to change these properties through Avenue scripts.

Assigning Access and Shortcut Keys

Menu options can be opened with the keyboard instead of the mouse if you define access keys. An access key is a selected character from an option's name. When the user holds the <Alt> key down and presses the access key, that menu option opens. Access keys are underlined when displayed on the menu bar.

Assign an access key by placing an ampersand before the desired character. For instance, *&Project* assigns *P* as the access key, so that pressing <Alt+P> opens the **Project** menu option. Access keys are not case sensitive and can be any printable character except the ampersand. Avoid using a space as the access key as it may interfere with the reserved keys for a windowing system. If you need the ampersand displayed, precede it with a backslash. For instance, *&Left\&Right* results in a menu label of *Left&Right* with *L* as the access key.

Menu items can also have access keys, and are assigned in the same manner as for menu options. Using menu item access keys, however, differs from using keys to open a menu option. Once the user opens the list of menu items, pressing the character assigned as the access key opens the chosen item.

Access keys may be duplicated across menu options or among menu items. However, only the first option or item assigned a particular access key is always selected. Duplicating access keys is not desirable in a well-designed application.

Certain operations that are used frequently, such as **Save** or **Tile**, merit direct access rather than always forcing the user to pass through respective menu options. You can provide direct access to any menu item by assigning a **Shortcut** property.

Select the desired menu item from the **Control Editor** area of the **Customize** dialog box. Double-click on the **Shortcut** property line inside the **Properties** list. The **Picker** dialog box, shown in the following illustration, appears with a list of available keys. Select the desired key and click the OK button. The selected key names will appear in the Shortcut property line. Users can execute that menu item by pressing the assigned shortcut key or key combination.

20   Customizing the Interface

The Picker dialog box with a list of keys.

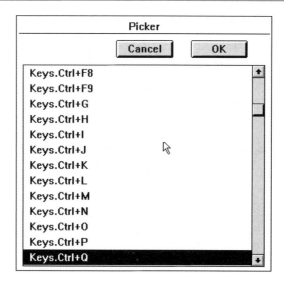

✗ **TIP:** Use <Shift> or <Ctrl> in shortcut key combinations to reduce accidental executions of a menu item.

Creating Pushbuttons

Activate the **Project** window by clicking the pointer on its title bar. Select the Project menu option from the Project window control bar to display a pull-down menu. From the pull-down menu, select the **Customize** menu item to display the Customize dialog box as shown below.

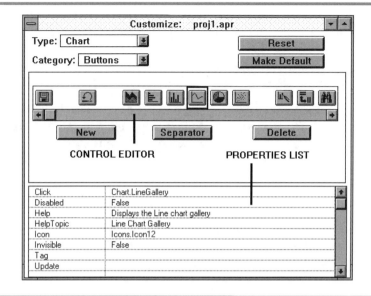

The Customize dialog box for a chart document.

Each ArcView document type has its own pushbutton bar. Select the pushbutton bar you want to customize by choosing the ArcView document from the **Type** option. The **Category** option must be set to **Buttons**. The current pushbutton specification is displayed in the **Control Editor** area.

Adding and Organizing Pushbuttons

Inside the **Control Editor** area, click on a pushbutton to select it. A frame appears around the selected button. To create a new button, click on the **New** button. The new button will be created to the right of the selected button. You can move a button by dragging it to a new location. To place a pushbutton at the beginning of the bar, create it, and move it to the beginning. If you drag the pushbutton and drop it on top of an existing one, ArcView places the new button to the left of the existing one. The Control Editor area is expanded by maximizing the **Customize** dialog box.

Although there is no limit on the number of pushbuttons you can create, ArcView displays only a single line of pushbuttons. Display

22   Customizing the Interface

resolution determines the number of buttons that can fit on the line. Too many pushbuttons may reduce readability and intuitiveness.

✘ **TIP:** Limit the number of your pushbuttons to 14, and never create more than 24.

You might consider using a separator between groups of pushbuttons to improve readability. To place a separator after a pushbutton, select that button and click on the **Separator** button.

ArcView creates pushbuttons with a blank default icon. Because pushbuttons cannot be labeled, it is important that the icon be representative of the button's action. To assign an icon to a pushbutton, select the button from the **Control Editor** area. Then double-click the pointer on the **Icon** line in the **Properties** list. The **Icon Manager** dialog box appears with a list of icons, shown in the following illustration. The list shows the icon name and shape. Select an icon and click on the OK button. The icon will appear in the Properties list and on the pushbutton. The purpose of the **Load...** pushbutton on the Icon Manager dialog box is for adding your own icons to the list.

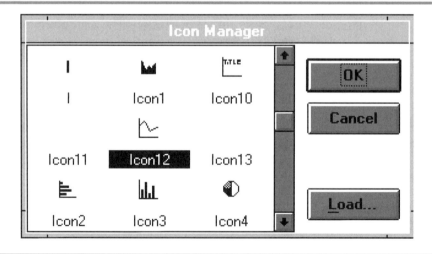

The Icon Manager dialog box with a list of icons.

Executing a Task Through a Pushbutton

A task can be executed through a pushbutton by associating a script to the button. You can use an existing system script, or write your own. Chapters 3 and 4 explain how to write new scripts.

Select the pushbutton to be affected from the **Control Editor** area of the **Customize** dialog box. Inside the **Properties** list, double-click on the **Click** property line. The **Script Manager** dialog box appears with a list of available scripts. The list includes both system and programmer-created scripts, as shown in the following illustration. Select the desired script from the list and click the pointer on the OK button. The selected script's name appears in the pushbutton Properties list. As discussed in Chapter 3, you can review the contents of system scripts or create new ones based on existing scripts.

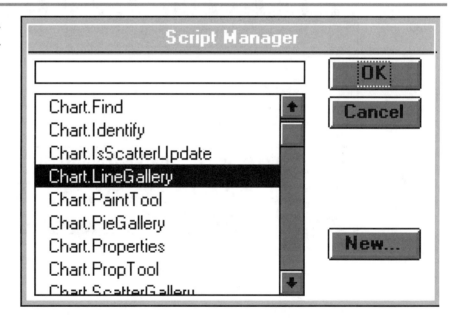

The Script Manager dialog box.

24   Customizing the Interface

Disabling and Hiding Pushbuttons

Under certain circumstances, your application may require that a specific pushbutton be unavailable for selection. You can either disable the button or make it invisible. A disabled button is displayed in gray but cannot be selected, while an invisible one is not displayed. Pushbuttons are not rearranged to fill in the gap caused by an invisible button. The following illustration shows the change in the display when the **Undo** button is made invisible.

Before and after making the Undo button invisible.

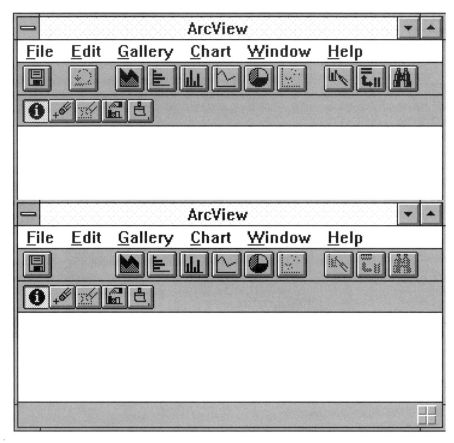

Select the pushbutton you wish to disable or make invisible from the **Control Editor** area of the **Customize** dialog box. The **Disabled** and **Invisible** property lines in the **Properties** list indicate the state of that button. A False value for the Disabled property means that the menu is not disabled and can be selected. A False value for the Invisible property means that the menu is displayed. Change the value of the Disabled or Invisible property by double-clicking its property line, which toggles between True and False. The "Event Programming" section in Chapter 4 explains how to change these properties through Avenue scripts.

Adding a Help Line to Pushbuttons

You can provide a short description, or help line, for every pushbutton. The help line will appear on the status bar whenever the user places the pointer on a button. This is a very helpful feature because users often forget the actions button icons represent. Take advantage of it. The following figure shows the location of the status bar.

The status bar.

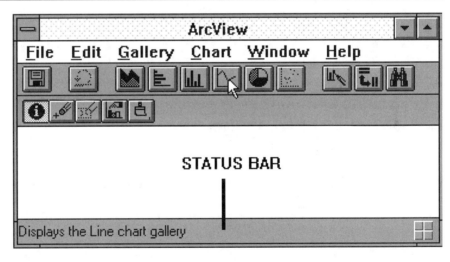

To define a help line, select the pushbutton to be affected from the **Control Editor** area. Double-click the **Help** property line inside the

26 Customizing the Interface

Properties list to display the Help dialog box, shown in the following illustration.

Although there is no limit on the number of characters, the help line should be short enough to be read quickly. If the help line extends beyond the status bar, it will be truncated. Display resolution and default font determine the number of characters that can fit on the status bar. Click the pointer on the OK button after typing the help line. The help line text will appear in the Properties list.

The Help dialog box.

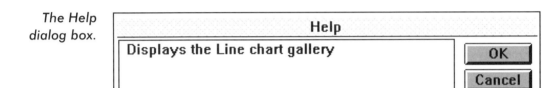

✘ **TIP:** The text for a pushbutton's help line should be a complete sentence, and preferably less than 14 words. Avoid using more than 24 words.

Creating Tool Buttons

To create a tool button, make the **Project** window active by clicking on its title bar. From the control bar of the Project window, select the Project menu option to display a pull-down menu. Select the **Customize** menu item from the pull-down menu to access the Customize dialog box.

Creating Tool Buttons 27

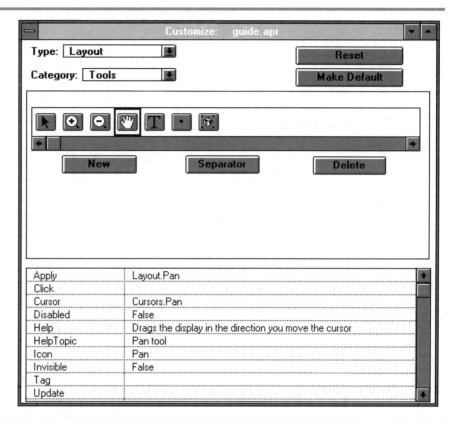

The Customize dialog box for a layout document.

Each ArcView document type has its own tool bar. Select the tool bar that you want to customize by choosing the ArcView document from the **Type** option. The **Category** option must be set to **Tools**. The current tool specification is displayed in the **Control Editor** area.

Adding and Organizing Tool Buttons

Select a tool from inside the **Control Editor** area. A frame will appear around the selected tool button. To create a new tool button, click on the **New** button. The new tool will be created to the right of the selected button. To place a tool button at the beginning of the bar, create it and then move it to the beginning. You can move a button by dragging it to its new location. If you drag the button and drop it on top of an existing

28   Customizing the Interface

one, ArcView places the new button to the left of the existing one. The Control Editor area is expanded by maximizing the **Customize** dialog box.

There is no limit on the number of tool buttons you can create, but ArcView displays all buttons in a single line. Display resolution determines the number of buttons that can fit on the line. Too many buttons could reduce readability and intuitiveness.

✘   **TIP:** Limit the number of tool buttons to 14, and never create more than 24.

ArcView creates buttons with a blank default icon. Since tool buttons cannot be labeled, it is important that the icon represent the button's action. To assign an icon to a button, select the button from the **Control Editor** area. Then double-click on the **Icon** line in the **Properties** list. The **Icon Manager** dialog box appears with a list of icons, as shown in the following figure. The list shows the icon name and shape. Select an icon and click on the OK button. The icon will appear in the Properties list and on the button.

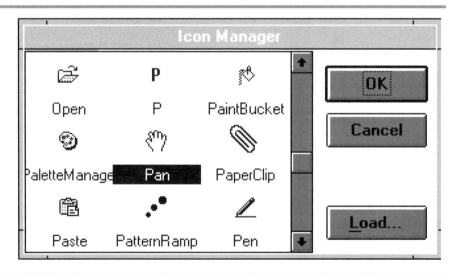

The Icon Manager dialog box with a list of icons.

✘ **TIP:** Because tool buttons are smaller than pushbuttons, an icon may be too large to fit on a tool button.

Executing a Task Sequence Through a Tool Button

A task sequence can be executed through a tool button by associating a script to the button. You can use an existing system script, or you can write your own. Chapters 3 and 4 explain how to write new scripts.

Select the button to be affected from the **Control Editor** area of the **Customize** dialog box. Inside the **Properties** list, double-click on the **Apply** property line to display the **Script Manager** dialog box. Available scripts are listed in this dialog box, as shown in the following illustration. The list includes both system and programmer-created scripts. Select the desired script from the list and click the pointer on the OK button. The name of the selected script appears in the tool button's Properties list. As discussed in Chapter 3, you can review the contents of system scripts or create new ones based on existing scripts.

The Script Manager dialog box.

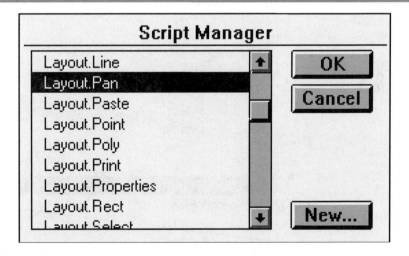

Associating a Cursor Shape to a Tool Button

When selecting a tool, the user expects to apply the tool's actions by clicking or dragging the pointer inside the document window. For instance, after selecting the **Zoomin** tool, the user expects to zoom into the area clicked on by the pointer. The user may repeat the tool's action as long as another tool is not selected. Consequently, it is important to change the shape of the cursor to indicate which tool is active.

You can associate a cursor shape to a tool button by selecting the button to be affected from the **Control Editor** area. Double-click on the **Cursor** property line inside the **Properties** list. The **Picker** dialog box appears with a list of cursors. The graphical shape and name of each available cursor are displayed in this list. Select the desired shape and click the OK button. The cursor name will appear in the Properties list.

The Picker dialog box with cursors.

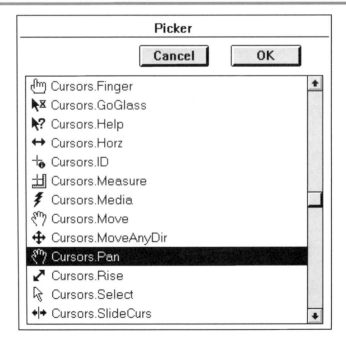

✘ **TIP:** To aid user recognition, the cursor's graphical shape should match or be conceptually related to the tool button's icon.

Disabling and Hiding a Tool Button

Your application may sometimes require that certain tool buttons be unavailable for selection. You can either disable tool buttons or make them invisible. A disabled button is displayed in gray to indicate that it cannot be selected. An invisible button is not displayed. Tool buttons are not rearranged to fill in the gap caused by an invisible button. The following figure shows the changes in the display when buttons are disabled and made invisible.

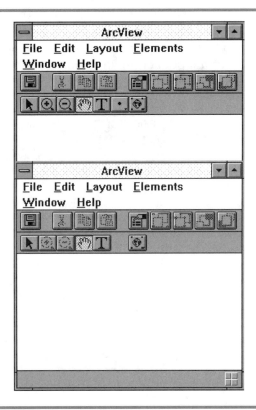

Before and after disabling and making invisible tool buttons.

32   Customizing the Interface

Select the tool button you wish to disable or make invisible from the **Control Editor** area of the **Customize** dialog box. The **Disabled** and **Invisible** property lines in the **Properties** list indicate the state of that button. A False value for the Disabled property means that the button is not disabled and can be selected. A False value for the Invisible property means that the button is displayed. Double-click on the Disabled or Invisible property lines to toggle between True and False. The "Event Programming" section in Chapter 4 explains how to change these properties through Avenue scripts.

Adding a Help Line to Tool Buttons

You should take advantage of the opportunity to provide a short description or help line for each tool button. This is a very helpful feature because users often forget which action a button's icon represents. The help line will appear on the status bar, shown below, whenever the user places the pointer on a button.

The status bar.

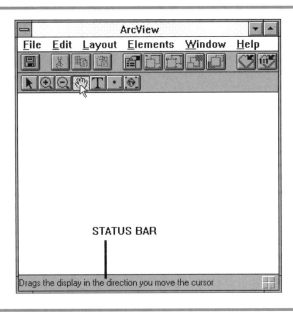

To define a help line, select the tool button to be affected from the **Control Editor** area. Double-click on the **Help** property line inside the **Properties** list. A help dialog box, shown in the following illustration, appears to accept the help line.

Although there is no limit on the number of characters in a help line, you should keep it short for easy reading. If the help line extends beyond the status bar, it will be truncated. Display resolution and default font determine the number of characters that can fit on the status bar. Click the pointer on the OK button after typing the help line. The help line text will appear in the **Properties** list.

The Help dialog box.

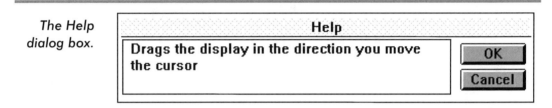

✗ **TIP:** Text for a tool button's help line should be a complete sentence and preferably less than 14 words. Avoid using more than 24 words.

Startup and Shutdown Scripts

Many applications can benefit from the ability to run a script based on the occurrences of an event, such as starting or ending a project. You can associate startup and shutdown scripts to your project. Startup scripts are useful for setup or verification of the application environment, while shutdown scripts are generally used for cleanup or validation.

Startup and shutdown scripts are attached in the **Properties** dialog box. The following illustration shows the Properties sheet for a project. Open the Properties dialog box by selecting the **Project** menu option. Choose the Properties menu item from the option's pull-down menu. Enter the script name in the Properties dialog box.

34   Customizing the Interface

*The Project
Properties
dialog box.*

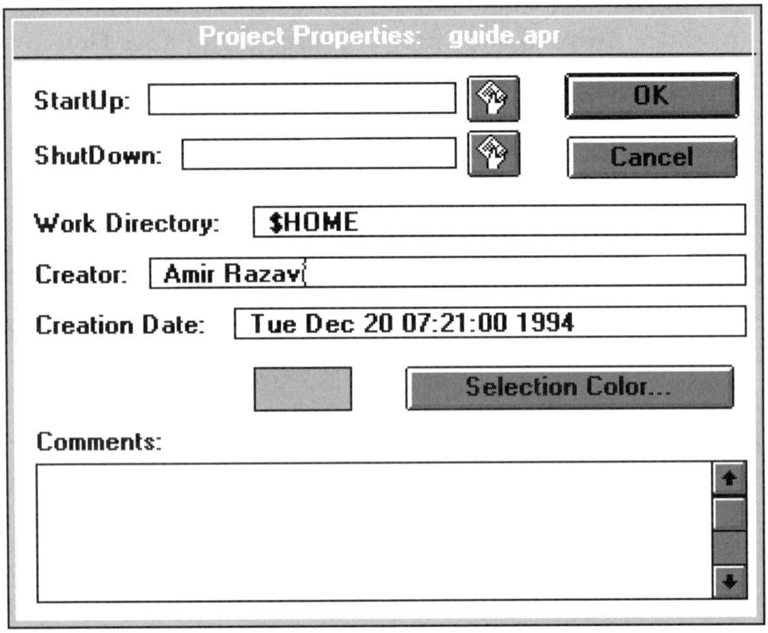

Project Properties: guide.apr

StartUp: [   ]

ShutDown: [   ]

Work Directory: | $HOME |

Creator: | Amir Razavi |

Creation Date: | Tue Dec 20 07:21:00 1994 |

Selection Color...

Comments:

OK

Cancel

You can use either a system script or a script that you have created. If you cannot recall the script's name, click the pointer on the **Loadscript** icon. The **Script Manager** dialog box appears with a list of available scripts from which to choose.

Saving the Customized Interface

Changes made to the interface are stored with the project. You can set the defaults so that the changes will appear with other newly created projects. ArcView has two default levels: system and user level. User level defaults supersede system level defaults. Defaults are stored in a project file named *default.apr*.

Save a customized interface by opening the **Customize** dialog box and clicking the pointer on the **Make Default** button. The Customize dialog box is accessed by making the **Project** window active, selecting the Project menu option, and then picking the Customize menu item. A user-level *default.apr* file is then created in one of the following platform-dependent directories:

- ❏ %HOME% for Microsoft Windows family
- ❏ The ArcView application folder for Macintosh
- ❏ $HOME for UNIX
- ❏ SYS$LOGIN for Open VMS

Do not overwrite the system default file, *default.apr*, in the *etc* directory, with the user level *default.apr*. The system default file contains all scripts and user interface controls while the user default file contains only the scripts and controls which are different from the system default. Chapter 12 explains how to create a system default project file.

✗ **TIP:** Existing projects are not affected by new defaults. If you press the **Reset** button in the **Customize** dialog box, your interface becomes the current system default setting.

More To Come

The user interface customization methods presented in this chapter are useful for preparing an interface in advance. As you learn Avenue programming language in the following chapters, you will also become familiar with how to customize the interface on the fly.

Chapter 3

Avenue Building Blocks

Welcome to Avenue. In this chapter, Avenue's characteristics and basic building blocks are presented, followed by a discussion of how to prepare and run Avenue scripts. You will soon discover that you can extend the flexibility and power of ArcView by using Avenue.

Avenue Characteristics

Avenue is ArcView's scripting language, and it can be used in the same fashion as AML (Arc Macro Language) is employed to create ARC/INFO applications. Avenue and AML are based on different development roots. While AML is a procedural language, Avenue is an object oriented programming language. Object oriented languages provide a higher degree of flexibility, stability, and reusability than procedural languages.

The three major characteristics of objected oriented programming languages are defined below.

Encapsulation

Procedural programming languages maintain data separate from procedures. In object oriented languages, data and procedures are kept together. An object consists of both data and code. Therefore, the paradigm for performing operations in an object oriented language is sending a "request" to an object that tells the object what we would like it to do. Requests are similar to function calls in procedural languages.

Polymorphism

The polymorphism characteristic of object oriented languages allows different objects to respond to the same message in their own unique ways. In most procedural languages, it is necessary to name functions differently when data types are different. For instance, both DivideInteger and DivideReal functions must be defined, even if each performs the same function of division.

In an object oriented language, only Divide (/) is defined, and the compiler selects the correct service based on the data type. Under polymorphism, the request will always be Divide, even though the code that provides this service may vary from class to class. This behavior provides a uniform interface within the program because the same name is used, while differences in code remain invisible to the programmer.

Inheritance

The inheritance characteristics of object oriented languages allow reuse of software by specializing already existing general solutions. Objects inherit variables and services from their parents. The parent class is also called the "superclass". When a request is sent to an object, the compiler searches for the corresponding service. If the compiler finds the corresponding service, the request is performed. Otherwise, the search process is moved to the object's superclass. This process continues to the top of the class hierarchy.

Some object oriented language compilers ignore unrecognized messages and others produce error messages. This behavior means that additional specialized functions can be created by adding the parts that are unique, while the rest are automatically inherited. For instance, the Enumerate class could incorporate the EnumerateDate and EnumerateAlphabet specialized classes. Services like Reset and Display are defined in the Enumerate superclass and inherited by the specialized classes. Services such as Increment and Decrement are defined within the specialized classes because their behaviors vary.

Basic Building Blocks

Class is the most basic concept in object oriented programming. A class provides the blueprint for creating objects, and is itself a type of object. Programs communicate with objects by passing messages, also known as requests in Avenue. If the request is recognized by the object, it provides a predefined service and returns a resultant object.

Classes

A class is a grouping of objects based on common characteristics. For instance, chart and table objects belong to the document class. In the initial phase of developing an application, you must identify the classes of objects that you need to create. After the first step of identifying the starting object, which is usually the project or the active document, object classes are identified, and then required objects are created. Avenue scripts in this book were developed according to this process. By examining these scripts, you can learn the required steps in developing Avenue programs.

Objects

An object oriented programming language creates an application by controlling objects. An object is defined as a person, place, or thing that knows the values of its attributes and can perform operations on those attributes. Objects are also known as "class instances", and operations

40   Avenue Building Blocks

are also called "methods" or "services". In ArcView, objects are the things that you work with, such as a line segment. Attributes for a line segment are the coordinates of the end points. The line object could provide a service that would move the position of one end.

Requests

Avenue controls an object by sending it a request, which can also carry parameters. The request is carried out by the object if the request has a corresponding service of the same name, and parameters required by the service are specified. Sending the same request to a different object can evoke different services. This characteristic is called polymorphism, as explained earlier.

An object inherits a service from its class or a superclass. Requests may also be made to a class, but such requests are generally limited to creating or obtaining information about class instances.

Interaction Between the Building Blocks

Avenue programs are an organized collection of requests made to classes and objects. New objects are created from classes, and existing objects are identified by ArcView. The request process is the interaction between Avenue building blocks.

Creating New Objects

To create an object, you must know its class. ArcView's class hierarchy, accessed through the help system, provides an efficient method for searching classes. Once you have become proficient in ArcView class hierarchy and definitions, it should be used as a verification tool. (See Appendix A for diagrams of the class hierarchy.)

Objects can also be created to hold existing data or to point to other objects. This process is similar to the concept of variables in procedural languages. Two examples of creating new objects follow.

Interaction Between the Building Blocks 41

```
myView = View.Make
myProjection = myView.GetProjection
```

The first line applies the **Make** request to the **View** class to create the *myView* object. In the second line, the **GetProjection** request to the object *myView* retrieves the projection from *myView*, and creates the object *myProjection* from that data. After these two lines are executed, *myView* references a view object, and *myProjection* references the projection object with the view *myView*. Any valid request to the *myView* and *myProjection* objects can now be made.

✔ **NOTE:** ArcView does not permit the creation of new classes or requests.

Making a Request

Avenue expressions are generally composed of (1) a variable which is a reference to an object, (2) an object, and (3) one or more requests. Requests are applied in one of three ways: postfix, infix, and prefix. The three examples below demonstrate each method.

```
myView.CopyThemes ' postfix
tomorrow = toDay + oneDay ' infix
not True ' prefix
```

The single quote character in Avenue represents a comment string, one of the language elements discussed in Chapter 4. In the first line, the postfix **CopyThemes** request is applied to an object with a period. The infix example shows how the **+** request sits between two objects. Finally, the prefix example places the **not** request before the object called *True*. ArcView's class hierarchy shows valid requests for an object and how they are applied. (See Appendix A for the Avenue class hierarchy.)

Requests can carry parameters required by the object. As shown in the following example, the parameters are passed along with the request by placing them in parentheses after the request.

```
MsgBox.Error(aMsg,aTitle)
```

42   Avenue Building Blocks

The ***Error*** request is asking the ***MsgBox*** class to create a window with an OK button and STOP icon. It also requests that the contents of the object *aMsg* be used as the error message, and that the contents of the object *aTitle* be placed in the window's title bar.

The ArcView class hierarchy can show details of a request's parameters. You must provide the same number and type of parameters in the order identified by the class hierarchy. In the example above, if you do not provide a title, ArcView uses the word "Error". This does not mean that only one parameter is passed to the object, but rather that the object *aTitle* does not have a string value. Use double quotation marks, as shown in the statement below, to indicate a null value. The following illustration shows the result of the statement.

```
MsgBox.Error("Unable to open theme","")
```

Example of an Error request to the MsgBox class.

Request names follow a convention that represents action and property. This convention is illustrated in the following examples.

```
aDisplay.GetUnits              ' Get returns a
                                 reference to an
                                 object's property.

myCircle.ReturnCenter          ' Returns a copy of an
                                 object's property.

aDisplay.SetUnits(newUnits)    ' Set applies a new
                                 value to an object's
                                 property.
```

Interaction Between the Building Blocks   43

```
myTheme.IsVisible              ' Is returns the state
                               of an object in Boolean
                               value.

myScript.HasErr                ' Has returns the state
                               of an object in Boolean
                               value.

myTheme.CanEdit                ' Can returns the
                               ability of an object in
                               Boolean value.

toDay.AsString                 ' As converts an object
                               into a new class.
```

The difference between the ***Get*** and ***Return*** requests is that Get points to the actual object while Return points to a copy of the object. Consequently, any changes to the object created with Get will change the original object.

Requests may be chained for code efficiency. ArcView processes chained requests from left to right. Chaining requests can reduce memory load by not setting variables. Avoid extensive chaining because it makes your script harder to read and more prone to errors. The following code segment shows an example of chained requests.

```
av.GetProject.FindDoc("Virginia").FindTheme("Counties"
).SetThreshold(aThreshold)
' The above statement is equivalent to the
' following code segment.
thisProject=av.GetProject
aView=thisProject.FindDoc("Virginia")
aTheme=aView.FindTheme("Counties")
aTheme.SetThreshold(aThreshold)
```

44   Avenue Building Blocks

Creating a Script

Avenue scripts are the code statements that collectively make up your application. If you are new to Avenue, start with simple scripts. Working out your script on paper first could make the process easier. You can then create your script by typing the code directly into the ArcView **Script Editor** or by using your own editor.

As you learn Avenue, or when scripts are small, you may find it easier to use ArcView's editor. However, once you formally start creating your application, it is advisable to use a system editor. There are two reasons for using a system editor instead of ArcView's. The primary reason is that ArcView stores scripts within the project file; by using a system editor you will have a text of your script to reuse. The second reason is that ArcView's editor is very limited in editing features; your system editor will probably be more efficient.

✘ **TIP:** Having access to ArcView's help system while writing scripts is useful. Open the help system and iconize it during sessions with your editor.

Using the ArcView Script Editor

The **Script** component provides an integrated development environment for developing Avenue programs. You can open this component by selecting the **Scripts** icon from the project window. At this point you can create a new script or open an existing one. Default names for new scripts are *Script1*, *Script2*, and so on. Always change the default to a more meaningful name by selecting the **Rename** item from the **Project** menu option. You can also rename a script in the **Script Properties** dialog box.

The Avenue programming environment, shown in the following figure, consists of the **Script Editor** window and its associated menu options and buttons.

The Avenue programming environment.

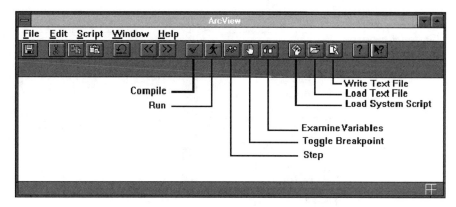

Scripts are saved within the project file rather than independently. Therefore, once you have finished writing a script, save it both as a script and a text file. You can save your code as a text file by clicking on the **Write Text File** pushbutton.

Using a System Editor

You can use any editor as long as the script is saved as a text file. Once you have completed the script, open a new **Script Editor** window by selecting the **Script** icon from the **Project** window, and then click on the **Load Text File** pushbutton.

A **Load Script** dialog box appears, shown in the next illustration. Select the text file and click on the OK button. ArcView then loads a copy of the file into its script editor. The loading process does not involve compiling or error checking.

The Load Script dialog box.

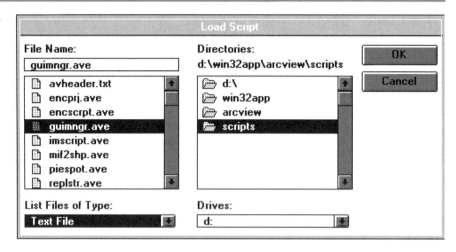

> ✘ **TIP:** When you make minor changes in the **Script Editor**, do not forget to update your original text file. Update the original file by saving the script as a text file.

Where to Start

When composing scripts, you need to access various objects to control or retrieve information. First, you need a starting object. For instance, to access a theme you first need the view object that is holding the theme. The most common method of retrieving the starting object is by finding the active object. Assume that you have developed a script to hide a theme's legend, and you have associated this script with a pushbutton for **View** documents. When the pushbutton is pressed, the first thing that your script should do is retrieve the view object. In the following statement, the variable called *theView* refers to the active document, in this case a view.

theView = av.GetActiveDoc

In the above statement, a request is sent to the object called *av*. The *av* object is important because it represents the executable ArcView. There is only one instance of *av*, and that is your current application.

Consequently, *av* is the starting object. Because your script is associated with a view control bar, the active document must be a **View**. The rest of the script will use the object called *theView* to access its themes.

✔ **NOTE:** An object is an instance of its respective class.

The statement below is another common way to start a script.

```
thisProject = av.GetProject
```

The object *thisProject* represents the current project. This approach is often used when the script is not associated to a specific document, window, or control bar. With the project object you can add or remove documents, or save the project.

Accessing System Scripts

A system script can be accessed for every user interface control object in ArcView. The script that you are writing may be a specialization or combination of some of these system scripts. If you believe that a particular system script's execution through a menu item, pushbutton, or tool button would be useful, go to the **Customize** dialog box and identify the script name. (On the access and use of the Customize dialog box, see Chapter 2.) Open a new script window and click on the **Load System Script** pushbutton to display the **Script Manager**. Scroll to the desired script, select it, and click on the OK button. The contents of the script are copied into your new script window. Repeat this process to insert other system scripts.

Script Testing

Avenue is a compiled language. Before using your script you must compile it and then test it for accuracy. While compiling and testing the script, Avenue tests for syntax and run-time errors.

Script Compiling

Before you run an Avenue script, the code must be compiled. Click on the **Compile** pushbutton to check for syntax errors. If no errors are found, ArcView compiles your script and activates the **Step** and **Run** buttons.

Your script will not run if it contains syntax errors, which are caused by failure to follow programming language rules. Simple typing errors are often identified by the compiler as syntax errors. If the compiler detects a syntax error, it displays a message and moves the cursor to the corresponding location in the script.

The compiler locates one error at a time. Therefore, if you correct an error and recognize other similar errors in your script, you will save time by correcting them before compiling again.

Start the **Script Editor** and key in the following code segment as a new script:

```
' This code segment has a syntax error.
thisProject=av.GetProject 'starting object
' Accessing a view document.
aView=thisProject.FindDoc("states")
if(nil=aView)
'Was the view found?
MsgBox.Error("States View not found","")
exit
end
```

Compile the script. The error message shown in the following illustration appears and the cursor moves to the location of the error. Can you correct this error?

An error message.

The corrected code follows:

```
' This code segment compiles without syntax error.
thisProject=av.GetProject 'starting object
' Accessing a view document.
aView=thisProject.FindDoc("states")
if(nil=aView) then '  Location of previous syntax
error.
'Was the view found?
MsgBox.Error("States View not found","")
exit
end
```

Another common cause of syntax error is the failure to define a variable or an object in the script. Avenue does not require declaration of variables at the start of a script. Variables are declared as references to objects when they are used on the left side of an assignment statement.

Running Your Script

You should test each script before associating it to a control object. Testing each piece of the application individually is known as "unit testing".

Click on the **Run** pushbutton to execute the script. It may be easier to unit test while an ArcView document is active. Create a temporary pushbutton for that document and associate your script to it. Then unit test your script by clicking on the temporary button.

Once you are satisfied that the script is running correctly, you can associate your script with a control object as explained in Chapter 2, or call it from another script as discussed in Chapter 4.

Fixing Run-Time Errors

Correcting run-time errors is more difficult. Some run-time errors will cause the script to crash during execution while others produce erroneous results.

50   Avenue Building Blocks

Use the **Run** button to execute your script. If the script stops before completion, a run-time error caused the crash. The following code segment shows how an invalid parameter can cause a programming crash.

```
' Code segment causing run-time crash.
thisProject=av.GetProject
aView=thisProject.FindDoc("MARYLAND")
if(nil=aView) then
if( MsgBox.YesNo("Create view?","",2) ) then
' Valid values for the third parameter of
' MsgBox.YesNo are True and False. The compiler
' does not catch this error.
newView=View.Make
newView.SetName("MARYLAND")
thisProject.Add(newView)
else
exit
end
```

Undefined classes or invalid requests can also halt the execution of a program. ArcView does not allow the creation of new classes or new services for existing classes. If you are not sure whether the class you have identified is defined, you should check the class hierarchy in the ArcView help system to verify the class you need and valid requests for the class.

These types of errors—invalid parameters, undefined classes, and invalid requests—can be located and resolved by using the **Step** button to step through the code.

Run-time errors that cause unexpected results are caused by program logic errors. The best approach for correcting this type of problem is a walk-through. Using the **Step** button, walk through the code. At critical points, use the **Examine Variables** button to review variable values.

Avenue Programming Language

To become fluent in any language you must first master basic elements such as syntax and grammar. This chapter presents Avenue's basic programming elements which include variables and control statements. If you have studied other programming languages, Avenue language elements will be familiar. Even if you are not a programmer, Avenue elements are easy to learn.

Basic operations that most scripts require are also covered in this chapter, such as communicating with the user, and accessing files. The chapter concludes with a section on a help system for your application.

Programming Elements

Most programming languages have common elements. If you are familiar with other programming languages, you can learn Avenue by recognizing

52   Avenue Programming Language

equivalent elements, such as the conditional If statement or the repeating While loop.

Referencing Objects with Variables

Your application can guide ArcView to execute many kinds of tasks such as displaying a map, opening the maps table, and a chart. To accomplish these and other tasks, the program needs information such as a view name or type of chart. These types of information are stored in "variables".

Variables in Avenue are references to objects. A variable name is a sequence of letters and digits starting with a letter. Although Avenue is not case sensitive, a common practice is to start a variable name with a lower case letter, and capitalize subsequent words. For example, *primeNumber, activeDocument,* and *vehicle* are all valid variable names. There is no limit to the number of characters in a variable name.

You should verify that variable names do not conflict with class names, or other reserved words. (See Appendix B for Avenue's reserved words.) The compiler produces an error message when a reserved word is used as a variable.

Next, a variable's scope is determined by its availability. For example, the scope of a local variable is limited to its script. The local variable exists only in its script; outside the script its value is not known. On the other hand, the scope of a global variable is the application. Once a global variable is created, its value is known to all scripts. Global variables are used to share objects between scripts. Any variable beginning with an underscore (_) becomes a global variable, e.g., *_symbolList*, and *_inputFileName*. When a script ends, the memory area allocated to local variables is returned to ArcView. Only the referencing variable is removed from memory, and not the object itself. In contrast, the memory area allocated to global variables must be cleared and returned to ArcView by issuing the following statement:

```
av.RemoveGlobalVars
```

Avenue does not require declaration of a variable before it is used. When you assign an object to a variable, Avenue initializes the variable. To assign an object to a variable, use one of the following formats:

```
variableName = object
variableName = expression
```

Avenue evaluates expressions and returns an object. If a variable appears on the right side of the equals sign before it is initialized, the compiler produces an error message. The following code lines show examples of variable assignment.

```
roadClass = "HEAVY"
unitReplacementCost = 5601.45
analysisBaseYear = Date.Make("1985","yyyy")
```

✘ **TIP:** Avoid meaningless variable names such as *a1* or *bb*. Program maintenance is easier when you use meaningful names, e.g., *landUseList*, or *roadwayTheme*.

Using Literals

Literals are objects created from the number, string, date, or time classes. Numbers can be floating (decimal), or integer. A string is a sequence of characters enclosed in double quotation marks. Examples of valid literals follow.

```
"Washington, D.C."
1001
3.14
07-04-1994
12:01:00
```

New string objects are created by assignment of a string literal, or by concatenation of two or more string objects. Avenue offers two concatenation operators: + and ++. The ++ operator inserts a single space between the two strings. String objects may contain non-printable characters such as a tab and newline. The **AsChar** request below converts a decimal value to its character form.

```
newLine=10.AsChar
```

You can also use the special object **NL** to produce a newline as shown in the following statements:

54   Avenue Programming Language

```
MsgBox.Info ("First message line" + NL +
"Second message line","")
```

String literals are also used to create other objects. An example of requests to convert objects appears in the statement below. These requests generally start with *As*.

```
Date.SetDefFormat ("MM-dd-yy")
deadLine = "06-01-94".AsDate
```

✗   **TIP:** Avenue evaluates chained mathematical requests from left to right unless parentheses are used to establish order of precedence.

The following code lines show examples of using string, date, and number objects.

```
city = "Manassas"
state = "Virginia"
label = city + "," ++ state
' The above label is "Manassas, Virginia".
'
theAnswer = 2 + 8 / 4 ' results in 2.5
theAnswer = 2 + ( 8 / 4 ) ' results in 4
'
dateBuilt = Date.Make ("1/1/65", "M/d/y")
today = Date.Now
```

Using Lists and Dictionaries

In object oriented languages, lists and dictionaries are also known as "collections". Collections are objects that contain other objects, and are useful in organizing related information. For instance, it is easier to keep the names of all your themes in a list called *allMyThemes* than scattered among many variables.

Avenue provides four classes of collections: lists, dictionaries, name dictionaries and stacks. Lists are series of objects or literals. Lists are created through direct assignment or by request. Stacks are similar to lists, except that objects are accessed in a last-in first-out method. Both lists and stacks can contain objects of different classes. The following code segment shows how lists are created.

```
roadClass = {"LIGHT", "MEDIUM", "HEAVY"}
bridgeClass = roadClass
bridgeClass.Add ("DAMAGED")
' The bridgeClass list is composed of LIGHT,
' MEDIUM, HEAVY and DAMAGED string objects.
'
roadAnnualCost = List.Make
roadAnnualCost.Add (25000)
roadAnnualCost.Insert (12000)
lightRoadAnnualCost = roadAnnualCost.Get (0)
' The variable lightRoadAnnualCost references
' the number object 12000.
```

Dictionaries are an efficient method to store and retrieve objects. In concept, they are similar to a two-column table with identifier and value column headings. Any type of object can be stored by a key, and the key may be from any class. Dictionaries accept objects and keys of different classes. In a name dictionary the key for an object is its name. The following code segment shows an example of how to use dictionaries.

```
roadTravelCost = Dictionary.Make (2)
' The (2) parameter refers to the size of
' hash tables and not number of objects in
' the dictionary. The number of objects should
' not exceed hash size times 20.
roadTravelCost.Add ("HEAVY",35.48)
roadTravelCost.Add ("MEDIUM",35.00)
roadTravelCost.Add ("LIGHT",34.28)
roadTravelCost.Add ("GRAVEL",35.50)
gravelRoadCost = roadTravelCost.Get("GRAVEL")
' The variable gravelRoadCost references
' the number object 35.50.
```

56   Avenue Programming Language

✘ **TIP:** Name dictionaries and other dictionaries are efficient for random access, while lists and stacks are efficient for sequential access.

Controlling Program Flow

Controlling the logic flow in any programming language is a basic operation. These controls range from executing a set of statements if a certain condition prevails, to executing the same set more than once as long as a particular condition exists. In all controls, the flow direction is based on a Boolean condition. The Boolean class includes True and False instances. The True and False objects are used for testing the condition of a control statement.

Boolean expressions are any sequence of requests resulting in True or False objects, and are used by the control statements to direct program flow. Compound expressions are created by using AND and OR requests. Examples of Boolean expressions appear below.

```
( themesList.Count > 0 )
' The above expression evaluates to True if
' the themesList list has any members.
'
( aStringList.FindByValue("STATE")>=0 AND
aStringList.FindByValue("COUNTY")>=0 )
' If both STATE and COUNTY string objects are
' members of aStringList, then it evaluates to True.
'
' The following Boolean expressions are true.
(True)
(NOT False)
'
' The following Boolean expression is false.
("ARC" = "VIEW")
```

Control statements are divided into conditional and loop structures.

✘ **TIP:** The *Exit* statement stops all running scripts.

Conditional Structure

An *If* statement is used to conditionally execute a series of statements. The condition is a Boolean expression resulting in True or False. For instance, you may want to display a theme's legend only if the theme is displayed.

The conditional statement is enclosed inside parentheses and may contain more than one Boolean expression. Use parentheses to establish precedence in evaluating compounded Boolean expressions. Otherwise, compound Boolean expressions are evaluated from left to right.

The expressions on both sides of an AND request must be true to obtain a True result. One or both expressions on either side of an OR request must be true to get a True result. The NOT operator reverses the expression result. The following code segment shows examples of If statements.

```
' This script displays State and County themes
' in an open View named USA.
thisProject = av.GetProject
thisView = thisProject.FindDoc("USA")
themesList = thisView.GetThemes
stateThemeNumber = -1
countyThemeNumber = -1
anIndex = 0
for each aTheme in themesList
if (aTheme.GetName = "State") then
stateThemeNumber = anIndex
elseif (aTheme.GetName = "County") then
countyThemeNumber = anIndex
end
anIndex = anIndex + 1
end
stateTheme = themesList.Get (stateThemeNumber)
countyTheme = themesList.Get (countyThemeNumber)
if (Not (countyTheme.IsVisible)) then
countyTheme.SetVisible (True)
```

58   Avenue Programming Language

```
end
if (Not (stateTheme.IsVisible)) then
stateTheme.SetVisible (True)
end
```

An *If* structure starts with the statement

```
if (condition) then
and it can have optional statements
elseif (conditional) then
```

or

```
else
```

and must end with the statement

```
end
```

If structures can be nested. A nested If structure is a conditional block inside another If block. If the condition is true, Avenue executes the program lines following the If statement until it reaches End, Else, or Elseif statements. All If structures end with an End statement. If the condition is false, Avenue moves to the first Else, Elseif or End statement. In an If structure, you can include many Elseif lines, but only one Else statement. Always use the Else statement after all Elseif statements. When Elseif and Else statements are part of an If structure, the Else block is executed only if none of the Elseif conditions are true.

✘ TIP: Whenever possible, avoid complex nested and compound If statements. They can be the source of run-time logic errors.

Loop Structure

Loop structures execute a set of code lines more than once. For example, you may write one set of codes to display a theme, and then iterate the statements and display all themes. Avenue offers ***While*** and ***For Each*** loops.

The structure of While loops is presented below.

```
While (condition)

code lines

End
```

If the condition is true, Avenue executes the statements inside the loop structure; otherwise, the program flows to the End statement. When Avenue encounters the loop, it first evaluates the condition, and if true, it executes the loop structure once, and returns to the top of the loop. Avenue evaluates the condition before executing every loop.

Loops can be nested. A nested loop is a loop inside another loop. Although Avenue has no limits on the number of nested loops, limits are imposed by computer memory size.

The following code segment shows an example of the **While** loop structure.

```
while (MsgBox.YesNo("Re-run the script?","", False))
av.Run("myScript",nil)
end
```

The **For Each** loop structure works with a list of values. The structure of For Each loops appears below.

```
For Each name in list

code lines

End
```

The list can contain several objects or a range of values. Avenue executes the code lines inside the loop structure for every value in the list. To skip list values, you can add the word "by" to the loop statement. The For Each loop can also be nested. The following code segment shows an example of the For Each loop structure.

```
' Assuming that for every bridge class in bridgeClass
' list the unit maintenance cost, average area, and
' life span are available in various dictionaries,
' this loop creates a dictionary for the annual
```

60 Avenue Programming Language

```
' maintenance cost.
bridgeMaintenanceCost = Dictionary.Make (2)
for each bridgeType in bridgeClass
cost = bridgeUnitCost.get(bridgeType) *
bridgeAvgArea.get(bridgeType) /
bridgeLifeSpan.get(bridgeType)
bridgeMaintenanceCost.Add(bridgeType,cost)
end
```

✘ **TIP:** When coding a loop structure, always verify how the loop ends. An endless loop will hang your application.

The program flow inside any type of loop structure can be altered by using Break and Continue statements.

A **_Break_** statement causes an immediate exit from the loop. If the Break statement is inside a nested loop, it affects only the innermost loop containing the statement. Break is often used when there is more than one reason to end the loop.

A **_Continue_** statement causes Avenue to skip the rest of the loop and go to the next iteration of the loop. Use the Continue statement to substitute for lengthy If statements in a loop structure. The code segment below shows an example of a Continue statement in a loop structure.

```
' Lets you copy up to five zone codes from a
' master list to a new list excluding the R1 zone.
newZone = List.Make
masterCount = masterZoneList.Count
for each index in 0..4
if (index >= masterCount) then
' The loop has passed the end of the master list.
exit
end

aZone = masterZoneList.Get(index)
if (aZone = "R1") then
' An R1 zone detected; skip the rest of the loop.
```

```
continue
end

newZone.Add(aZone)
end
```

Documenting with Comment Lines

Documenting your program can save time in the future when you need to maintain the application code. The Avenue character for comments is a single quote ('). Wherever this token appears, everything to the end of the line is read as a comment. The exception is when a single quote appears inside a string. The following code segment shows examples of a comment token placed at the beginning and the middle of a line, and an instance of where the token does not precede a comment.

```
' This comment starts at the beginning of a line.
theView = theProject.GetActiveDoc ' Comments go here.
MsgBox.Info("I miss the 70's","")
' The single quote in the preceding statement does
' not create a comment line.
```

✘ **TIP:** Never procrastinate about documenting your code. Chances are that you will not do it later. Always start your program with comment lines containing your name, date, program name, version number and purpose.

Event Programming

The term "event" refers to the occurrence of an action. For instance, when the user makes a theme active, an event has occurred inside the view document. Task initiation based on an event is called "event programming". User interface controls have **Click**, **Update**, and **Apply** event attributes. An Avenue script can be associated to these attributes, and the scripts are executed when a click, update or apply event takes place.

62   Avenue Programming Language

A click event occurs when a user selects the menu item or pushbutton. An apply event occurs when a user clicks on the mouse in the display area while a tool button is active. Click and Apply events were explained in Chapter 2. Update events occur when the state of a document changes. For example, opening a document or selecting a feature triggers an update event.

Update event programming is useful in a variety of situations, although it is used mainly for enabling or disabling user interface items. For instance, you might create a tool button that computes bridge replacement cost in a road coverage. If the theme for the road coverage is not present, the button should be invisible. If the theme is present but not visible, the tool button should be disabled. Finally, the tool button should be available for use only when the road coverage is displayed.

An Avenue script for update event programming is associated with the Update property of a user interface item. The script consists of two parts: testing for an event and performing a task. ArcView automatically executes the script every time an update event occurs. The following sections describe how to test for different events.

Checking for a Condition

Requests with an *Is* prefix are used in checking for a condition. For example, you can use the *IsAtEnd* request to determine whether you have reached the end of a file. Several of Avenue's numerous requests starting with *Is* are listed below.

```
aTheme.IsActive
aChart.IsChartScatter
Coverage.IsINFO (myCoverage)
aControl.IsEnable
aString.IsNumber
aTool.IsSelected
aTheme.IsVisible
```

The above *Is* requests return a True or False Boolean object. The following code segment shows how a tool button is enabled when a theme is visible.

```
' This script is associated with the
' update property of the tool button.
if (theTheme.IsVisible) then
Self.SetEnabled (True)
else
Self.SetEnabled (False)
end
```

Checking for Active Objects

Requests with a **GetActive** prefix are used in accessing active objects. An object is made active or inactive by clicking the mouse on the object. Active objects are generally displayed differently; for instance, an active theme is raised while an active field is lowered. These requests return the active object(s), or nil if there are none. A short list of **GetActive** requests follows.

```
av.GetActiveDoc
aTable.GetActiveField
aView.getActiveThemes
```

The following code segment shows how selecting a numeric field enables a customized pushbutton.

```
' This script is associated with a pushbutton
' for statistical analysis on the selected
' numeric field.
aField = theTable.GetActiveField
if (nil = aField) then
Self.SetEnabled (False)
elseif ( aField.IsTypeNumber ) then
Self.SetEnabled (True)
else
Self.SetEnabled (False)
end
```

64 Avenue Programming Language

Interaction Between Programs

If your application is small and narrow in scope, a single Avenue script may be sufficient. Most applications, however, require interaction among several Avenue scripts, and skillful developers design applications in modular form. In a modular design, application functions are broken down into separate Avenue scripts. Maintainability and reusability are among the major benefits of modular design.

Modular design, however, introduces a new challenge to the application developer: how will the program modules interact? The two principal types of interaction are calling other programs and sharing information.

In a typical modular design, a master script controls the entire application and calls upon other scripts to carry out specific functions. The master script is often called the "main program", and the scripts it calls are known as "subroutines". Subroutines perform specific tasks using information supplied by the main program. The next two sections explain how the main program calls its subroutines and shares information with them.

Calling Other Programs

An Avenue script can start another script by using the **Run** request. The format of such a request follows.

```
av.Run (scriptName, ownerObject)
```

Avenue searches first for the script identified by **scriptName** in the current ArcView project file. If the script is not found in the current project, Avenue searches the user and system default projects. Avenue displays an error message if the script is not found. Once the subroutine has completed its task, Avenue returns to the next statement in the main program. The following code segment is an example of the **Run** request.

```
' Present a list of possible projections.
prjList = "Lambert Albers Azimuthal".AsList
whichPrj = MsgBox.ChoiceAsString
(prjList,"Select a projection: ","")
```

```
' Stop the program if user clicks on
' the cancel button.
if (nil = whichPrj) then
exit
end

' Call appropriate subroutine to
' perform the projection.
if (whichPrj = "LAMBERT") then
av.Run ("LAMBERT_PRJ","")
elseif (whichPrj = "ALBERS") then
av.Run ("ALBERS_PRJ","")
else
av.Run ("AZIMUTHAL_PRJ","")
end
```

If you plan to use a script from the user or system defaults, you should first verify that the script exists. If the script you need is not available, you can prevent your application from crashing by using the **FindScript** request to carry out a search. An example appears below.

```
aScript = av.FindScript ("ALBERS_PRJ")
if (nil = aScript) then
MsgBox.Error
("Unable to find the script.","")
exit
else
av.Run ("ALBERS_PRJ","")
end
```

If the script is located, **FindScript** returns the script name. FindScript returns nil if the script is not located.

✘ **TIP:** In a "top-down" approach to application development, you begin by writing the main program to call subroutines that perform various functions. The main program could be your application's main menu. Under this approach, each function is called based on the user's selection from the main menu.

Avenue Programming Language

Passing Objects Between Programs

A subroutine often requires information from the main program to complete its task. For instance, the *LAMBERT_PRJ* subroutine mentioned in the previous section requires parameters such as central meridian, reference latitude, and standard parallels for a Lambert Conformal Conic projection. The main program will also need to know if the projection was carried out successfully when *LAMBERT_PRJ* ends.

The ***ownerObject*** in the ***Run*** request (see "Calling Other Programs" in the previous section) provides the means to share information between the main and subroutine programs. By using the keyword ***Self***, a subroutine references its ***ownerObject***. Storing shared data in a dictionary and using the dictionary as the ***ownerObject*** is a useful practice. The following code segments illustrate this method. The first block appears in the main program, and the second block appears in the subroutine.

```
' Store predefined projection parameters
' in the main routine.
sharedDictionary = Dictionary.Make (2)
sharedDictionary.Add ("CM",cenMeridian)
sharedDictionary.Add ("RL",refLatitude)
sharedDictionary.Add ("P1",firstParallel)
sharedDictionary.Add ("P2",secondParallel)
sharedDictionary.Add ("SUBJECT",theView)
sharedDictionary.Add ("EXTENT",theExtent)
sharedDictionary.Add ("SUCCESS",False)
' Run the subroutine that performs the projection.
av.Run("LAMBERT_PRJ",sharedDictionary)
' Check to see if projection was completed.
if (Not (sharedDictionary.Get("SUCCESS"))) then
MsgBox.Warning
("Unable to perform the projection","")
end

' In the subroutine create and set the projection;
' refer to sharedDictionary as SELF.
```

Accessing Files 67

```
thisLambert = Lambert.Make(Self.Get("EXTENT"))
thisLambert.SetCentralMeridian (Self.Get("CM"))
thisLambert.SetReferenceLatitude (Self.Get("RL"))
thisLambert.SetLowerStandardParallel (Self.Get("P1"),
thisLambert.setUpperStandardParallel (Self.Get("P2"))
thisLambert.Recalculate
thisView = Self.Get("SUBJECT")
thisView.SetProjection (thisLambert)
Self.Remove ("SUCCESS")
Self.Add ("SUCCESS",True)
```

✘ **TIP:** You do not have to develop all subroutines before testing the main routine. Create a skeleton with a message box to show that a particular subroutine was called. This process is similar to prototyping, or walking the user through your application without real-time task execution.

Values can be returned to the calling script by using the ***Return*** statement. If you do not specify a Return statement, the last object created in the subroutine is returned. An example of the Return statement appears below.

```
' In the main program:
worked = av.Run ("LAMBERT_PRJ",sharedDictionary)
' In the subroutine:
return True
```

Accessing Files

Application files can be sources of input, temporary storage areas, output destinations, or all three. Avenue recognizes text and line files, and both types store printable characters. Therefore, the number 123 is stored as the three characters, 1, 2, and 3. Generally, each character corresponds to one byte, except in some non-English versions of an operating system. The difference between text and line files is in how Avenue reads and

68   Avenue Programming Language

writes their contents. Avenue accesses a text file one character at a time, while line files are accessed one line at a time. A line is a series of characters ending with a platform-dependent delimiter. (In Microsoft Windows, the delimiter is comprised of both a carriage return and newline character; in Macintosh, carriage return character; and in the UNIX environment, the newline character.) The length of each line is limited only by memory size.

To read a file or write into it, you must first open or create the file by using a file name object. The file name object and associated tasks are described in the next section.

The File Name Object

Avenue stores file names as objects of the ***FileName*** class. A thorough understanding of this class is needed for using requests that handle file objects. ***FileName*** is an operating system-independent class that, through its instances, refers to valid file names in various operating systems. The file may or may not physically exist.

Before using any file, a FileName class object should be created. The ***Make*** request, as shown below, creates a ***FileName*** object.

```
myFilename = FileName.Make ("POP90.DAT")
```

In this example, *myFilename* becomes a file name object, and *POP90.DAT* is the object's value. Requests to the ***FileName*** class also deal with the file path. In the following example a file name is created with the system's temporary directory as its path.

```
tmpDirName = FileName.GetTmpDir
myFilename = FileName.Make ("AVGDATA.TMP")
myFilename = FileName.Merge (tmpDirName, myFilename)
```

An alternative is to replace the last two lines of the above example with the following two lines:

```
myFilename = FileName.Merge (tmpDirName,
"AVGDATA.TMP".AsFilename)
```

If you know the temporary path, all three lines can be replaced with the single statement below:

```
myFilename =
FileName.Make("C:\Windows\temp\AVGDATA.TMP")
```

In multi-user applications, unique names for each user's temporary files must be assigned to prevent confusion. To create a unique name, you can use the **MakeTmp** request as illustrated below.

```
' A prefix of up to five characters and an extension
' can be established for unique file names.
aUniqueFilename = tmpDirName.MakeTmp ("AHR","TMP")
```

The first string is the prefix, and the second string is the extension for a unique file name. A prefix or extension could identify the application that has generated the file.

Creating, Opening, and Closing a File

The **Make** request to the **TextFile** or **LineFile** classes creates new files or opens existing files. When a new file is created, it is also opened. The following examples illustrate opening or files.

```
' Create a text file to write to and read from;
' if file exists, overwrite it.
myTextFile = TextFile.Make
(myTextFilename, #FILE_PERM_WRITE)
' Open a line file for reading.
myLineFile = LineFile.Make
(myLineFilename, FILE_PERM_READ)
```

The first parameter for the Make request is an instance of the **FileName** class, which assigns the file name. The second parameter establishes the file permissions. These permissions are listed below.

- ❏ *#FILE_PERM_READ* opens an existing file for reading.
- ❏ *#FILE_PERM_WRITE* creates a new file for writing. If the file exists, it is overwritten.
- ❏ *#FILE_PERM_MODIFY* opens an existing file for reading and writing. You must use the **Flush** request after each write, and the **SetPos** request before each read.

70   Avenue Programming Language

❏ *#FILE_PERM_APPEND* opens an existing file for appending. If the file does not exist, a new file is created.

❏ *#FILE_PERM_CLEARMODIFY* opens a new file for reading and writing. If the file exists, it is overwritten. You must use the ***Flush*** request after each write, and the ***SetPos*** request before each read.

A file object created in this manner can accept requests to read from or write to. Avenue generates an error message if a file access conflicts with file permissions.

Open files may be damaged if the system crashes. Therefore, you should open a file only when it is needed and close it as soon as it is no longer needed in the application. Because opening and closing files are time-consuming, you should avoid unnecessary calls.

Once a file is closed it cannot be used again until reopened. To close a file, make the ***Close*** request to your file object, as shown below. Avenue closes any open file object at the end of your application.

```
anyFile.Close
```

✗   **TIP:** Multiple instances of a file can be opened by several file objects that reference the same physical file. Each file object must be closed once the operation is completed.

You can delete a file or check for its existence by making ***Exists*** and ***Delete*** requests to the ***File*** class. Each request, as shown below, needs a file name as its parameter.

```
' Delete an old file named OLD_DATA.
oldFilename = File.Make ("OLD_DATA")
foundIt = File.Exists (oldFilename)
' Exists request returns True or False;
' True means that the file exists.
if (foundIt) then
File.Delete (oldFilename)
else
MsgBox.Info (oldFilename.AsString++
"was not found","")
end
```

Accessing Files **71**

Reading and Writing a File

Before listing the requests that carry out reading and writing operations, a brief discussion of "file pointers" and "file elements" is in order. A file pointer points to a location inside a file. Read and write operations start at the position of the file pointer. When a file is created or opened, the file pointer is positioned at the beginning unless the *#FILE_PERM_AP-PEND* permission is used. In that case, the pointer is positioned at the end. Avenue moves the pointer by sending the ***SetPos*** request.

A file element refers to the quantity of data Avenue reads or writes per request. The element size is based on file type: one character for text files, and one line for line files.

Read and ***Write*** are basic requests that apply to all file types. The format for these requests appears below.

```
aFileObject.Read (dataBuffer, elementCount)
aFileObject.Write (dataBuffer, elementCount)
```

The ***elementCount*** is a numeric object referring to the number of elements to read from the file into the ***dataBuffer*** or to write from the dataBuffer into the file. The dataBuffer is made up of a list of file element objects. For instance, the dataBuffer for a line file is a list of lines.

The following statements illustrate how a line file accepts requests to read and write complete lines:

```
aLine = aLineFile.ReadElt
aLineFile.WriteElt (aLine)
```

The first statement reads the next line and stores it in the *aLine* string object. The second statement writes the characters in the *aLine* object to the *aLineFile*.

After each read or write, Avenue positions the file pointer at the end of the data just accessed. As illustrated in the code segment below, the pointer's position can be changed by using requests that handle the pointer's position.

```
' Place the pointer at the beginning of the file.
aFile.GotoBeg
```

72   Avenue Programming Language

```
' Place the pointer at the end of the file.
aFile.GotoEnd

' Check to see if we are at the end of the file.
atTheEnd = aFile.IsAtEnd
```

Two other requests allow you to determine the position of the pointer or the file size.

The *GetPos* request returns the sequence number of the element the pointer is pointing at. Elements in a file are numbered from zero. Consequently, if the *GetPos* request returns 1, the pointer is pointing at the beginning of the second element.

As demonstrated below, the *GetSize* request returns an integer which shows the number of elements in a file.

```
elementCount = myLineFile.GetSize
MsgBox.Info ("There are" ++ elementCount.AsString ++
"Lines in the file","")
```

User Dialog

The primary method of communication with the user is through the message box class called *MsgBox*. All requests are made to the class rather than to class instances. An object cannot be created from this class.

All message boxes are modal: the user must respond to the box before continuing. Message boxes are used to provide a message or accept an input. The action is based on the request.

Displaying Information, Warnings, and Errors

Message display requests are listed below:

```
MsgBox.Info (aMessage, aTitle)
MsgBox.Warning (aMessage, aTitle)
MsgBox.Error (aMessage, aTitle)
```

User Dialog 73

In the preceding statements, the *aMessage* and *aTitle* objects are strings that will display as part of the message box. An error, warning, or information icon is displayed in the box. The box will also contain an OK pushbutton for user acknowledgment. The next figure shows the warning message box displayed by executing the following statement:

MsgBox.Warning ("Unable to find the theme.","")

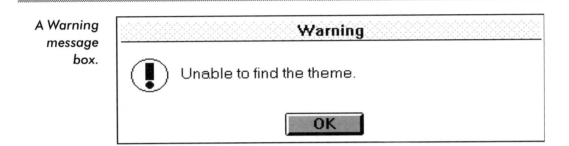

A Warning message box.

✗ **TIP:** You can omit the title by keying in two double quotation marks. Do not use nil if you intend to omit the title.

Getting an Input

The following requests are used to solicit a Yes or No answer from the user:

MsgBox.YesNo (aMessage, aTitle, defaultFlag)
MsgBox.MiniYesNo (aMessage, defaultFlag)
MsgBox.YesNoCancel (aMessage, aTitle, defaultFlag)

The *aMessage* and *aTitle* parameters are string objects that will be displayed in the message box. Each message box will contain YES and NO pushbuttons. The **YesNoCancel** request creates a message box which includes a CANCEL pushbutton. The statement returns a True or False object. If the YES button is clicked, the statement returns True. If the NO button is clicked, the statement returns False. A nil is returned when the CANCEL button is pressed. The *defaultFlag* is a Boolean object that assigns a default pushbutton to the box. A value of False selects NO as the default button, and a value of True assigns YES as the default button. Users can select the default pushbutton by pressing <Enter>.

The next illustration shows a message box created by the following code segment:

```
saveIt = MsgBox.YesNoCancel
("Save project before exiting? ","",1)
if (saveIt) then
thisProject.Save
exit
elseif (Not saveIt)
exit
end
```

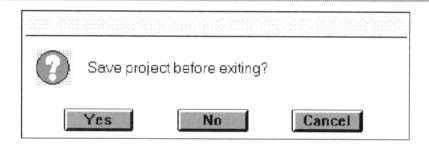

A YesNoCancel message box.

The following series of requests are used to ask for data input:

```
inputString = MsgBox.Input (aMessage,
aTitle, defaultInput)
inputString = MsgBox.Password
inputObject = MsgBox.Choice (aList, aMessage, aTitle)
inputObject = MsgBox.ChoiceAsString (aList,
aMessage, aTitle)
```

The **Input** request provides an input line, an OK button, and a CANCEL button in a message box. If the user presses the CANCEL button, this request returns nil; otherwise, the string entered by the user is returned. The **defaultInput** parameter establishes a default string for the input box. If you don't have a default value, enter two double quotation marks.

The **Password** request creates a message box similar to the **Input** request except that asterisks are displayed instead of the characters keyed in by the user.

The **Choice** request creates a message box with a drop-down list of objects from the *aList* parameter. The **GetName** request is applied to each object of the list. If the object name is not meaningful to the user, use the **ChoiceAsString** request, which applies the **AsString** request to each object from the drop-down list and press the OK button to return the selected object

✘   **TIP:** Use the **GetName** request for objects which have a name property, such as themes. For objects lacking a name property, such as a date object, use the **AsString** request.

Use the **MultiInput** request if you wish to accept more than one piece of data per message box.

Getting a File Name

By making requests to the **FileDialog** class, you can have the application ask the user for file names. The FileDialog class can accept **Put** and **Show** requests. Because **FileWin** has no class instances, an object cannot be created from this class. The result is a modal window that lists directories and existing files, and allows disk navigation. The window also has an input area for the file name, an OK pushbutton, and a CANCEL pushbutton. Examples of requests to the FileDialog class are shown in the following statements:

```
newFilename = FileWin.Put
(defaultFilename, namePattern, aTitle)
existingFilename = FileWin.Show
(namePattern, patternLabel, aTitle)
```

Avenue Programming Language

The *defaultFilename* is a **FileName** class object which appears in the input box of the window, but it can be replaced with a different name. The *namePattern* is a string object that determines the file names listed in the window. For instance, the *.* string lists all files, and *.APR lists files with the *APR* extension. The *patternLabel* identifies the file type for the *namePattern*. If the *namePattern* is *.DBF, then the *patternLabel* can be "DATABASE FILES". The *aTitle* is a string object appearing in the window title bar.

The **Put** request requires input of a new file name. If the file name exists, Avenue asks the user to confirm replacement of the existing file. The only input the **Show** request permits is existing files. The following statement creates the window shown in the next figure.

```
testFilename = FileDialog.Show ("*.HLP",
"Help Files", "")
```

A Filename window.

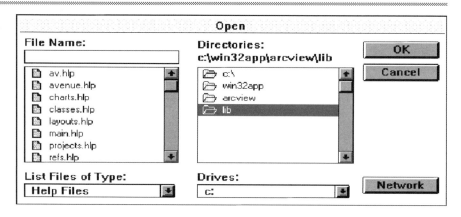

✘ **TIP:** Changing directories through the **FileDialog** window also changes the current working directory.

Upcoming Events

Chapters 3 and 4 present sufficient information to develop small or simple applications. In the following chapters, various aspects of Avenue are explored in depth. Examples provide a hands-on approach to learning how to develop more complicated applications.

Chapter 5

Programming the Project Window

A collection of associated documents in ArcView is called a "project". The project window organizes and displays the names of the documents. Although your application will generally access the associated documents more frequently than the project window, the project window has a set of features that can be useful in all applications.

In this chapter we will take a high-level look at an application for estimating damages caused by a brush fire. When working with this application, the user loads coverages of planimetric features along with the final boundary of the damaged area to estimate total damages to property and infrastructure.

The chapter ends with a tutorial that walks you through the process of creating a simple application.

Setting Up an Application

How might typical users start your application? They could double-click on an icon or type the application name on a command line. Whatever the process you choose, it must be easy to remember and simple to perform. This section presents an alternative, shown in the following illustration, that modifies ArcView's interface with menu items to start an application.

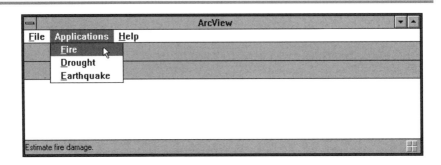

A customized ArcView interface.

A Customized ArcView User Interface

The purpose of the above interface is to provide the user with a quick and convenient way to begin an application. Listed below are the steps required in customizing the interface as discussed in Chapter 2.

❐ Start a new project.

❐ Make the **Project** window active and select the **Customize** menu item from the **Project** menu option.

❐ Set the category option to **Menu,** and the type option to **Appl** (application).

❐ Make invisible the **New Project** menu item in the **File** menu options.

✘ **TIP:** When you don't need a control interface such as a menu option or tool button, make it invisible instead of deleting it. You may need it later.

Create a menu item for each of your applications. The scripts initiate your applications by opening designated projects. For instance, the following script starts the *Fire* application. The next figure shows the properties of the Fire menu item. For each menu item, you must create an Avenue script. Once everything is in place, click on the **Make Default** button and close the **Customize** dialog box.

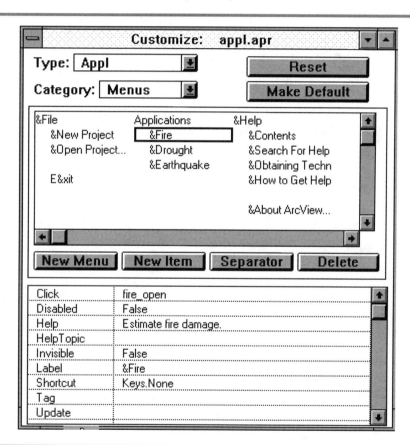

The Fire menu item properties.

The following script starts a project named *_FIRE*. The purpose of the underbar is to create a unique file name; it is unlikely that users will begin project file names with an underbar. Users should save projects with new names. For example, assume that on three occasions a user

82 Programming the Project Window

starts the _FIRE_ application through starting the project of the same name. The application is activated to estimate the damage from a specific fire, and results are saved in projects named _CALAV_, _SHASTA_, and _ELDORADO_, respectively. The _FIRE_ project file remains free of data and can be used at any time for a new fire, while _CALAV_, _SHASTA_, and _ELDORADO_ hold information on individual fires.

```
' Start the _FIRE project.
projectFile = FileName.Make ("_FIRE.APR")
Project.Open (projectFile)
```

Once a specific project file such as _SHASTA_ is created, the user will open that project directly from the **File** menu option. For instance, if the user wants to update the _SHASTA_ project, ArcView would be started, and the **Open Project** menu item selected from the **File** option. The user would then select _SHASTA.APR_ from the file window.

✔ **NOTE**: ArcView opens an empty project each time it is started. In order to see the **Applications** menu option you just created, you must close the empty project.

The Application Startup Script

The best time to enforce saving the project with a new name is when ArcView opens the _FIRE_ project. The following Avenue script should be placed in the **Startup** property of the _FIRE_ application to force a save with a new name.

```
' Forcing Save As at the startup.
theProject = av.GetProject
defaultName = FileName.GetCWD.MakeTmp ("FIRE","APR")
' Open the file window to get a project name.
newProject = FileDialog.Put
(defaultName,"*.APR","Create a New Project")
if (nil <> newProject) then
theProject.SetFileName(newProject)
theProject.Save
else
```

```
' If user clicked on Cancel, close the project.
theProject.Close
end
```

Setting Project Parameters

The **Project Properties** dialog box provides the means to assign Avenue scripts to start up and shut down events. To open this dialog box, select the **Properties** menu item from the **Project** menu option. The following figure shows the Project Properties for the _FIRE application.

The Project Properties dialog box.

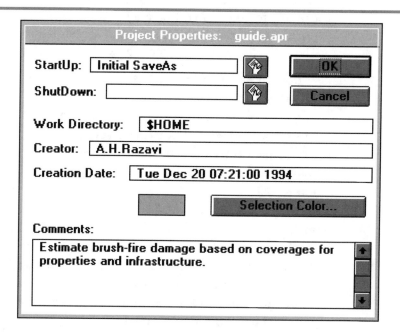

This dialog box can also store general information for the project. The importance of documentation cannot be overemphasized, so key in your name in the **Creator** field and provide a short description under **Comments**.

Finding a Document

The project window provides direct access to all documents associated with the project. The ArcView user can create, open, or delete documents with the project window. An application developer can use the project object in Avenue to access document objects. For instance, if you want a list of themes to appear in a view, the view object must be accessed. To access the view object, you can search for it in the current project object.

There are two ways to reference a document object. When the Avenue script is associated with a document through an interface control, you can ask for the active document. If the script is not associated with a document in this fashion, you can search for the document by name. The following code segment shows how to reference a document.

```
' Open the view named Final.
thisProject = av.GetProject
theView = thisProject.FindDoc ("Final")
theViewWindow = theView.GetWin
if (theViewWindow.IsOpen) then
theViewWindow.Activate
else
theViewWindow.Open
end
' Show or hide the legend for the current view.
' Associate this script with a pushbutton.
theView = Av.GetActiveDoc
for each oneTheme in theView.GetThemes
oneTheme.SetLegendVisible
(Not(oneTheme.IsLegendVisible))
end
```

Arranging Document Windows and Icons

During the course of an ArcView session, users could create or open many documents. At some point, the ArcView window becomes cluttered. The **Window** menu option can be used to manually organize all documents, arranging them as tiles or cascading them next to each other. Another option is to click on their minimize buttons to reduce them to icons. These capabilities are also available for inclusion in your application through the following Avenue statements.

```
av.TileWindows
av.CascadeWindows
av.ArrangeIcons
```

In the following code segment, _FIRE_ opens several views and then arranges them in tile format.

```
' Open views of fire.
view3amWin.Open
view9amWin.Open
view3pmWin.Open
' Arrange views on the screen.
av.TileWindows
```

Displaying a Help Message

Window-based applications with graphical user interfaces (GUIs) rely on the user to direct the application in executing functions. This reliance on the user is the tradeoff for the flexibility offered by windows.

In contrast, a non-GUI application requests pertinent data from the user in a serial fashion. Failure to respond to some questions is allowed, but a non-GUI application will not start computation until all questions have been displayed. If the same application was equipped with a GUI, the user could indicate which data is available for input by selecting menu items or pushbuttons, thereby avoiding the display of irrelevant questions. In addition, the user could click on a pushbutton at any time to start computation based on the available data.

In a GUI application, a short message indicating how to start, or what the next step is, can be a great help to the user. For instance, if _FIRE_ expects selection of a base coverage or an image file, the following message could be displayed on the status line:

```
Load a base map or image file by selecting the Load
menu option.
```

86 Programming the Project Window

The Avenue requests that handle messages appear below.

```
av.ShowMsg ("Message string")
av.ClearMsg
```

Messages appear in the status line. The status line is also used by ArcView to display help lines for the interface items. When incorporating help messages into your application, keep in mind that ArcView messages are also displayed in the status line. Thus, if your application and ArcView act to display messages at the same moment, application-generated messages will be overridden by ArcView's.

One effective use of status line messages is to confirm process or task completion. In this case, the advantage of using the ***ShowMsg*** request over a message box request (***MsgBox***) is that ShowMsg does not require acknowledgment. For the same reason, however, you should not use the ShowMsg request to display critical information.

There is no limit on the number of characters in a status line message. ArcView, however, truncates messages longer than the status line. The + and ++ operators can be used to assemble the message string.

Displaying the Status Bar

You may have noticed that when ArcView opens a project, it displays a status bar showing the progress of opening the project. A progress report is useful for application tasks that are usually time-consuming, and lack other means to show progress, such as changes to a view. Another benefit of using a status bar is that the user can interrupt the process.

The Avenue requests which show a status bar and stop button appear below.

```
av.SetStatus (aNumber)
av.ShowStopButton
```

The parameter *aNumber* ranges from 0 to 100. The status bar is displayed at zero, and at 100 the status bar is cleared. Each ***SetStatus*** request moves the status bar to the location indicated by *aNumber*. The request returns a Boolean True value unless the user has clicked on the stop button.

Setting Up an Application   87

The following code segment is a simple example of how the status bar can be used to show a progress report.

```
' Start the status bar with stop button.
av.ShowMsg ("Computing damages...")
canceled = False
av.ShowStopButton
statusIndex = 0
av.SetStatus (statusIndex)
' Status bar reaches 100% when computation
' is performed for all themes.
themesCount = themesList.Count
statusIncrement = 100 / themesCount
' Application loops through all themes.
for each aTheme in themesList
av.Run("compute_damage",aTheme)
statusIndex = statusIndex + statusIncrement
continued = Av.SetStatus (statusIndex)
if (Not continued) then
canceled = True
break
end
end

' Handle any interruption.
if (canceled) then
av.ShowMsg ("Process interrupted.")
else
av.ShowMsg ("Process completed.")
end
```

88 Programming the Project Window

Tutorial: Creating an Application

This tutorial gives you a taste of Avenue programming by guiding you through the development of a simple application. The tutorial assumes that you are knowledgeable about ArcView, and do not need instruction on tasks such as saving a project or renaming a view.

For the practice application, we will use the *USA* coverage shipped with ArcView. You should be able to locate the coverage in ArcView's *avdata\namerica* directory. The application provides menu options to select one or more states, and enables the user to build a population density chart, print the chart data to a file, and compose a map.

Development Overview

When designing window-based applications you must consider the following questions: Which objects are needed by the application? How does the application respond to events?

You will work with ArcView objects, such as menu items, charts, or graphic elements. Events are actions taken by the user, such as clicking the mouse on a pushbutton or opening a view document. The application responds to events through the Avenue scripts you write. The combination of ArcView objects and Avenue scripts make up your application.

The following objects are required for the application tutorial:

❑ Menu options and items

❑ A view document with the State theme

❑ The theme's attribute table

❑ A chart document

❑ A layout document

The Menu Items

The menu items used here are Select States, View Chart, Compose Map, and Report.

Tutorial: Creating an Application   **89**

Select States. When this menu item is selected, the application will open the view document with the *State* theme. The program verifies that the view extent is set to the theme, and clears any other selected state.

View Chart. Selecting this menu item opens a predefined chart document and redraws its contents to match the current selection. This menu item must be disabled if no state has been selected.

Compose Map. This menu item opens a layout document to allow the user to compose a map of states.

Report. This menu item generates a delimited text file from selected records. If no records are selected, all records are written.

The Development Stages

The practice application is created in the following stages:

❑ **Stage 1.** Create required documents such as the view or the chart.

❑ **Stage 2.** Customize the interface by adding the menu option and items.

❑ **Stage 3.** Develop Avenue scripts that respond to the selection of menu items.

Stage 1: The Application Documents

In this section, you will create a project file and a new view document, and join two tables.

1. Create the project file by opening a new project and saving it as *GUIDE.*

2. Create a new view document and rename it *USA_View.* Access the **Properties** dialog box of this document to rename it.

3. Load the *State* theme into the *USA_View* document.

4. Make the *State* theme active, and from the **Theme** menu option select the **Table** menu item. This action loads and opens the *Attributes of State* table.

5. Close the table and view documents.

90   Programming the Project Window

6. Click on the **Table** icon in the project window. Load the *Stdemog.dbf* table by selecting the **Project** menu option and **Add Table...** menu item.

7. Join the *Stdemog* table containing demographic information with the *Attributes of State* table. The result of the join should go to the latter table. Join the two by opening both tables and selecting the *State_name* field in each table. Then, while the *Attributes of State* table is active, select the **Join** menu item from the **Table** menu option.

8. Keep the *Attributes of State* table open and active. Create a chart by selecting the **Table** menu option and **Chart** menu item. In the **Chart Properties** dialog box, add *Pop90_sqmi* to the **Groups** list and place 0IState_name in the **Label series using** field.

9. Rename the newly created chart *Density_Chart*.

10. Save the project.

Stage 2: Customize the Interface

In this section, you will customize the interface using techniques discussed in Chapter 2.

11. Open the **Customize** dialog box by double-clicking on the pushbutton bar.

12. Set the **Type** option to **Project** and the **Category** option to **Menus**.

13. Click on the **New Menu** pushbutton to create a new menu option. Verify that the new menu option is selected. Change its name by double-clicking on the **Label** property. When the **Label** dialog box appears, key in *&Guide* and click on the OK pushbutton.

14. Add a menu item to the **Guide** menu option by clicking the mouse on the **New Item** pushbutton. Change the name of the new item to *&Select States* by double-clicking on its **Label** property. Add shortcut keys by double-clicking on the **Shortcut** property to display the **Picker** dialog box. From the **Picker** dialog box, select **Keys.Shift+F1** and click on OK. Finally, add a short help text.

Tutorial: Creating an Application 91

Double-click on the **Help** property to open the **Help** dialog box. In the **Help** dialog box, key in *Select states from the USA View document*; click on OK. The following figure shows the **Customize** dialog box.

The Customize dialog box with the Select States menu item.

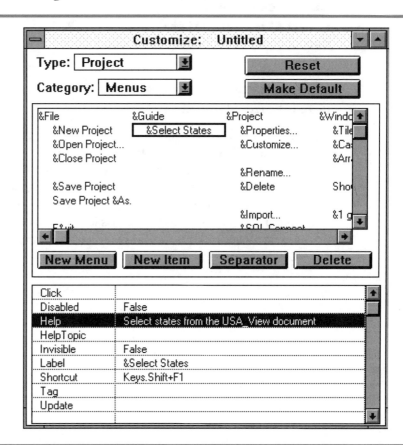

15. Use the following specifications when adding three additional menu items to the **Guide** menu option. See the instructions in step 4 above.

92   Programming the Project Window

```
Label: View &Chart
Shortcut: Keys.Shift+F2
Help: View density chart for the selected states

Label: Compose &Map
Shortcut: Keys.Shift+F3
Help: Open a layout document

Label: &Report...
Shortcut: Keys.Shift+F4
Help: Print data to a file
```

16. In the **Customize** dialog box, change the **Type** option to **View**. Maintain the **Category** at **Menus**. Add the **Guide** menu option with **View Chart** and **Compose Map** menu items according to steps 4 and 5 above. The following figure shows the **Customize** dialog box for this step. The user interface for view documents has a **Graphics** menu option with G as the access key. Change the **Graphics** access key to lower case r to avoid confusion with the G access key in the **Guide** menu option.

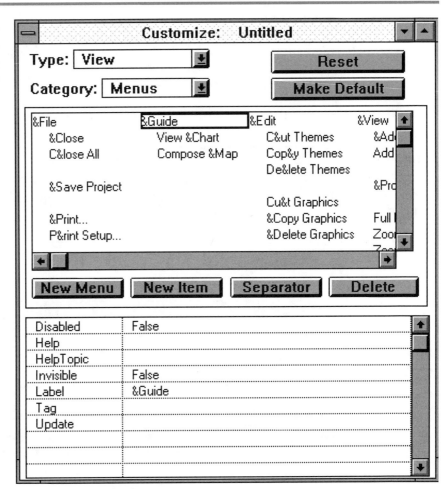

The Customize dialog box for View Document.

17. In the **Customize** dialog box, maintain the **Category** option at **Menus** but change the **Type** option to **Chart**. Add the **Guide** menu option with **Select States**, **Compose Map**, and **Report...** menu items according to steps 4 and 5.

18. The user interface for chart documents has a **Gallery** menu option with *G* as the access key. To avoid any confusion with the **Guide**

94   Programming the Project Window

menu option access key, change the **Gallery** option access key to lower case *y*. To change the access key, verify that the **Type** option is **Chart**, select the **Gallery** menu option, and double-click on the **Label** property. When the **Label** dialog box appears, change *&Gallery* to *Galler&y*.

19. Compare your user interface to the next three figures.

The Project Window User Interface.

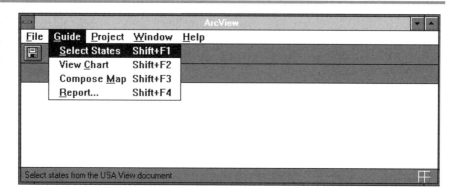

The View Document User Interface.

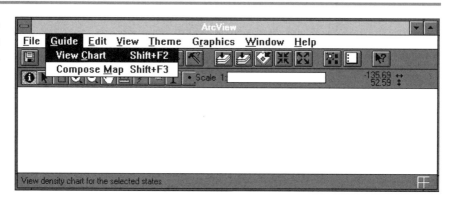

The Chart Document User Interface.

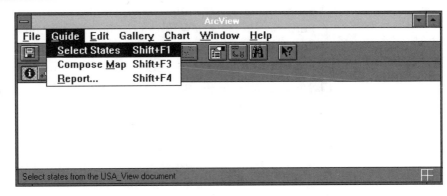

20. Save the project.

Stage 3: Avenue Scripts

Thus far, you have assembled the objects required by the practice application, as well as the necessary ArcView documents and user interface controls. The next phase is the design of the application's response to events. The events for this application are the user selections of menu items you just created. Responses are the actions that your application takes when the menu items are selected. For instance, when the user selects the **View Chart** menu item, the application should open the *Density_Chart* document you created earlier.

You will need the following five Avenue scripts.

❏ Script name: *Guide Select States*
Purpose: Open the USA_View document.

❏ Script name: *Guide Update Chart*
Purpose: Disable the View Chart menu item if no state is selected.

❏ Script name: *Guide View Chart*
Purpose: Open the Density_Chart document.

❏ Script name: *Guide Print*
Purpose: Write the selected records to a comma-delimited text file.

96   Programming the Project Window

❏ Script name: *Guide New Layout*
Purpose: Create a new layout document or open an existing layout document.

The project name, *Guide*, precedes all script names for two reasons:

❏ When you open the **Script Manager** dialog box to associate these scripts to respective menu items, menu items will be listed in groups by script, and thus, easy to find.

❏ As you write numerous scripts in the future, including the project name as part of the script name can be used as a script management tool.

A script is created by opening the **Script Editor**. In the **Project** window, click on the **Script** icon, and select the **New** pushbutton to open the **Script Editor**. ArcView gives each new script a default script name, which is not usually descriptive of the script's contents. You should rename the new script with a more meaningful name. A script is renamed by opening its property dialog box. Select the **Script** menu option and then select the **Properties** menu item to open the properties dialog box.

The Guide Select State Script

Key in the following Avenue code and name it *Guide Select State*. Click on the **Compile** pushbutton. If there are no syntax errors, save your work. If the compiler locates errors, review your script for any deviation from the code.

```
' Guide Select State
' Open the USA_View document.
'
guideProject = av.GetProject
usaView = guideProject.FindDoc("USA_View")
'
' Stop if the view document does not exist.
if (nil = usaView) then
MsgBox.Error ("USA_View document does not
exist.","Guide")
```

Tutorial: Creating an Application   97

```
exit
end
'
' Open the view document.
usaViewWin = usaView.GetWin
if (usaViewWin.IsOpen.Not) then
usaViewWin.Open
else
usaViewWin.Activate
end
'
' Get the state theme's object.
stateTheme = usaView.FindTheme ("State")
'
' Stop if theme does not exist.
if (nil = stateTheme) then
MsgBox.Error ("State theme does not exist.","Guide")
exit
end
'
' Display the theme.
if (stateTheme.IsVisible.Not) then
stateTheme.SetVisible (True)
end
'
' Clear any selected features.
stateTable = stateTheme.GetFTab
allSelected = stateTable.GetSelection
allSelected.ClearAll
stateTable.SetSelection(allSelected)
'
' Zoom to the extent of state theme.
if (stateTheme.IsActive.Not) then
```

98   Programming the Project Window

```
stateTheme.SetActive (True)
end

usaView.GetDisplay.SetExtent
(stateTheme.GetExtent.Scale(1.1))
```

Associate this script with the **Select State** menu item. Remember that the Select State menu item appears in more than one place. Open the **Customize** dialog box and click on the **Select State** menu item. Double-click the item's **Click** property to open the **Script Manager** dialog box, and select the *Guide Select State* script from the list.

The Guide Update Chart Script

Use the following code to create a new script called *Guide Update Chart*. Compile the code and then associate it with the **Update** property of the **View Chart** menu item. Remember that this menu item appears in more than one place.

```
' Guide Update Chart
' The View Chart menu item should be disabled
' when no state has been selected.
'
guideProject = av.GetProject
usaView = guideProject.FindDoc ("USA_View")
'
' Keep menu item disabled if view does not exist.
if (nil = usaView) then
Self.SetEnabled (False)
exit
end

'
' Count selected feature items.
stateTable = usaView.FindTheme("State").GetFTab
selectedCount = stateTable.GetSelection.Count
'
' Disable or enable based on the count.
```

Tutorial: Creating an Application 99

```
if (selectedCount > 0) then
Self.SetEnabled (True)
else
Self.SetEnabled (False)
end
```

The Guide View Chart Script

The following script opens the *Density_Chart* document and redraws its contents. Name the script *Guide View Chart*, compile it, and associate it with all the **View Chart** menu items.

```
' Guide View Chart
' Open Chart named Density_Chart.
'
guideProject = av.GetProject
densityChart = guideProject.FindDoc ("Density_Chart")
'
' Stop if the document does not exists.
if (nil = densityChart) then
MsgBox.Error ("Density_Chart document does not
exist","Guide")
exit
end

'
' Open the document.
densityChartWin = densityChart.GetWin
if (densityChartWin.IsOpen.Not) then
densityChartWin.Open
else
densityChartWin.Activate
end

'
' Redraw the chart to accept new selection.
densityChartWin.Invalidate
```

100   Programming the Project Window

The Guide Print Script

Associate the following script with the **Report** menu item. This script asks the user to supply a file name and then writes the selected records to the file.

```
' Guide Print
' Write the selected records to a delimited
' text file, and if none are selected write all
' records.
'
' First, get the table object.
usaProject = av.GetProject
usaTable=usaProject.FindDoc ("Attributes of State")
if (nil = usaTable) then
MsgBox.Error ("Attributes of State table does not
exist",
"Guide")
exit
end

'
' Get an output file name.
aFileName=FileDialog.Put ("".AsFileName,"*.*","Print
to a File")
if (nil = aFileName) then
' Stop if the user clicked on the Cancel button.
exit
end

'
' Write to the file.
usaVTab=usaTable.GetVTab
usaVTabSelectionCount = usaVTab.GetSelection.Count
if (usaVTabSelectionCount = 0) then
usaVTab.Export(aFileName,Dtext,false)
MsgBox.Info("All states written
```

Tutorial: Creating an Application   101

```
to"++aFileName.GetFullName,
"Guide")
else
usaVTab.Export(aFileName,Dtext,true)
MsgBox.Info("Selected states written
to"++aFileName.GetFullName,
"Guide")
end
```

The Guide New Layout Script

The script below requests the user for a layout name. If the layout exists, the script opens it; if the layout name does not exist, the script creates the new layout. Associate this script with the **Compose Map** menu item.

```
' Guide New Layout
' Create a new layout document
' or open an existing one.
'
usaProject = av.GetProject
'
' Get a layout name from the user.
layoutName = MsgBox.Input ("Enter a layout name:
","Guide","")
if (nil = layoutName) then
' Stop if the user clicked on the Cancel button
' or did not provide a name.
exit
end

aLayout = usaProject.FindDoc (layoutName)
if (nil = aLayout) then
' Create a new layout.
aLayout = Layout.Make
aLayWin = aLayout.GetWin
aLayWin.Open
```

102   Programming the Project Window

```
aLayout.SetName (layoutName)
usaProject.AddDoc(aLayout)
else
' Open the existing layout.
aLayWin = aLayout.GetWin
aLayWin.Open
end
```

Testing Your Application

The testing process for large applications often includes a test plan and document to guarantee a complete check. For small applications such as this one, a formal plan may not be necessary. However, rigorous testing is always recommended regardless of application size, and you should consider asking others to test your program. Software bugs can be quite embarrassing.

Begin the testing process by using your program under normal conditions. Unfortunately, many developers stop here. Careful planning and design generally preclude mishaps when running the application under normal conditions. However, the causes of application crashes are special circumstances. You should actually try to crash your application by avoiding standard steps and procedures. For instance, what happens if the *USA_View* document is accidentally deleted by a user? What happens if a user creates several layouts with the same name?

Of course, these "what-ifs" can be escalated to an absurd level. It is up to you to decide how bullet-proof your application should be.

Your Learning Curve

In this chapter we discussed several features common to all ArcView applications. The tutorial provided you with an opportunity to experience the thought processes and stages of development involved in creating a simple application. The most efficient method of gaining experience in designing ArcView applications is a gradual increase in the complexity of your work. Try to avoid skipping stages in your learning curve by attempting to create a large and complex application before its time.

In Chapters 6 through 11 specialized features are examined by concentrating on each type of ArcView document.

Programming View Documents

In this chapter we will create a simplified crime analysis application. The exercise includes creating a view document, and adding text and other graphical elements.

Following an overview of the application, the Avenue scripts are developed. Next, the customized interface required for the application is reviewed. (See Chapter 2 for details on developing customized interfaces.)

Application Overview

The application to be developed here consists of a base map and the following series of Avenue scripts.

CRM_Get_Locations accepts a crime incident point from the view document through clicking of the mouse. The script then asks the user for the point label. A graphical point symbol and label text appear on the view document.

- *CRM_Analysis* draws a circle which incorporates all incident points. At least three points should be present. Lines are drawn from the center of this circle to each incident point.

The scripts described below make the application "friendlier".

- *CRM_Scale1* and *CRM_Scale2* change the extent of the view display to one of two pre-set values. One extent is at the neighborhood level, and the other is at the regional level.
- *CRM_New_Analysis* clears all graphical elements and resets all variables for a new set of incident points.
- *CRM_New_View* creates a new view document and loads a pre-set base map.

The customized interface consists of a tool button for the *CRM_Get_Locations* script, and pushbuttons for the remaining scripts. The pushbutton for *CRM_New_View* is placed in the project window user interface. The other control buttons are placed in the view document user interface, as shown in the following illustration.

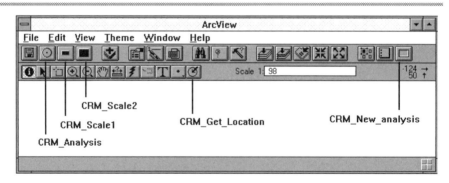

The customized user interface for view documents.

The View Document

The first script to be developed is *CRM_New_View*. In this script you create a new view document, set its properties and scale units, open its window, load a theme, and zoom to the extent of that theme.

```
' CRM_New_View
'
' Create a view document.
thisProject = av.GetProject
newView = View.Make
if (nil = newView) then
MsgBox.Error ("Unable to create a new view", "CRM")
Exit
end

'
' Get a View name and set properties.
newViewName = MsgBox.Input ("Enter a new view
name:", "CRM", "")
if (nil = newViewName) then
MsgBox.Warning ("Program stopped by the user", "CRM")
Exit
end

newView.SetName ( newViewName )
newViewDpy = newView.GetDisplay
newViewDpy.SetUnits ( #UNITS_LINEAR_FEET )
'
' Display the document.
newViewWin = newView.GetWin
newViewWin.Open
'
' Load a theme.
areaMapName = SrcName.Make ("WARD_3")
areaMapTheme = Theme.Make (areaMapName)
```

108 Programming View Documents

```
newView.AddTheme (areaMapTheme)
'
' Zoom to the view's extent.
newViewDpy.SetExtent(newView.ReturnExtent.Scale(1.1))
```

Two other scripts, *CRM_Scale1* and *CRM_Scale2*, are also developed in the initial phase. Because the only difference between the two is the scale value, only one script is developed here. You can easily develop the second one on your own.

```
' CRM_Scale1
'
' Set the extent of the current
' View to the neighborhood.
thisView = av.GetActiveDoc
thisDpy = thisView.GetDisplay
thisDpy.SetExtent(thisView.ReturnExtent.Scale(0.5))
```

Creating a New View

The scripts in your application typically act upon an existing view prepared by the user. However, some applications could include predefined views created by scripts. Creating a view document is accomplished by sending a **Make** request to the **View** class. This request also adds the returned object to the project object. The following code segment illustrates this step.

```
thisProject = av.GetProject
newView = View.Make
if (nil = newView) then
MsgBox.Error ("Unable to create a new view", "CRM")
Exit
end
```

Displaying a View Document

To open an existing document you need access to the document's window object. Once you have access, you can send an **Open** request to display it, as shown in the following code segment.

```
thisProject = av.GetProject
theView = thisProject.FindDoc ("View1")
theViewWin = theView.GetWin
if (theViewWin.IsOpen.Not) then
theViewWin.Open
end
```

Setting View Properties

Avenue allows you to change the properties and parameters of a view. You can generally find a specific request to send to the view object to set a certain parameter. Appearing below is a list of requests for setting parameters and properties, such as name and projection.

```
aView.SetProjection (aProjection)
aView.SetTOCWidth (aLength)
aView.SetName (aName)
aView.SetComments (commentString)
```

The following code segment sets the name and map units for the view.

```
newViewName = MsgBox.Input ("Enter a new view
name:", "CRM", "")
if (nil = newViewName) then
MsgBox.Warning ("Program stopped by the user", "CRM")
Exit
end

newView.SetName ( newViewName )
newView.GetDisplay.SetUnits ( #UNITS_LINEAR_FEET )
```

110   Programming View Documents

Setting the Display Extent

Avenue provides you with requests that change the extent of a displayed map. An application that uses multiple views to simultaneously display different parts of a map requires this type of request. The following code segment sets the extent to display all themes.

```
theView.GettDisplay.SetExtent
(theView.ReturnExtent.Scale(1.1))
```

Other requests are available through the view's display object. In the following code segment, Avenue sets the display extent to the active themes.

```
themesList = theView.GetActiveThemes
aBox = Rect.MakeEmpty
for each oneTheme in themesList
aBox = aBox.UnionWith (oneTheme.GetExtent)
end
theView.GetDisplay.SetExtent (aBox.Scale(1.1))
```

Graphical Elements

In this section, the *CRM_Get_Locations*, *CRM_Analysis*, and *CRM_New_Analysis* scripts are developed. The *CRM_Get_Locations* script is associated with a tool button and executes at the click of a mouse. The next figure shows the properties of the tool button that runs this script. The script places a point symbol at the coordinates where you click the mouse, and stores the points in a list attached to the view so that *CRM_Analysis* can access them. By attaching the list of points to the view document, the user can run *CRM_Analysis* at any time.

Graphical Elements 111

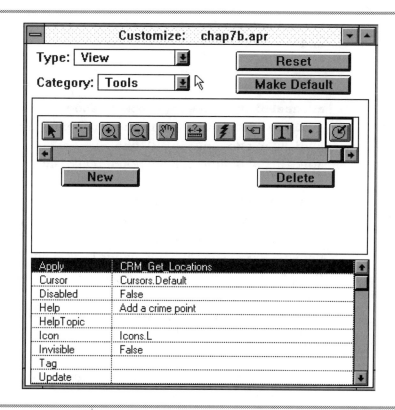

Tool button properties for the CRM_Get_Locations script.

```
' CRM_Get_Locations
'
' Get the display and graphic list for this view.
theView = Av.GetActiveDoc
theDpy = theView.GetDisplay
theGList = theView.GetGraphics
'
' Get the list object attached to the view,
' or create it if it does not exist.
pointList = theView.GetObjectTag
if (nil = pointList) then
pointList = List.Make
```

112   Programming View Documents

```
theView.SetObjectTag (pointList)
end

'
' Get a point.
aMousePoint = theDpy.GetMouseLoc
' Clone the mouse point to preserve its value.
aPoint = aMousePoint.Clone
aGPoint = GraphicShape.Make (aPoint)
'
' Get a label for this point.
aLabel = MsgBox.Input ("Point label:", "CRM", "")
if (nil = aLabel) then
MsgBox.Warning ("Point canceled", "CRM")
Exit
else
aGLabel = GraphicText.Make (aLabel,aPoint)
end

'
' Add the point to the list and display.
theGList.Add (aGPoint)
theGList.Add (aGLabel)
pointList.Add (aPoint)
```

The *CRM_Analysis* script below uses the list of points attached to the view to determine a bounding rectangle for all points. The script then draws a circle through the four corners of the rectangle. Lines are also drawn from the center of the circle to each point on the list.

```
' CRM_Analysis
'
theView = Av.GetActiveDoc
pointList = theView.GetObjectTag
if (nil = pointList) then
MsgBox.Warning("There are no points to
analyze","CRM")
```

Graphical Elements 113

```
Exit
else
theDpy = theView.GetDisplay
theGList = theView.GetGraphics
end

'
' Must have at least three points to analyze.
if (pointList.Count < 3) then
MsgBox.Error ("A minimum of three points is
required","CRM")
exit
end

' Establish the bounding rectangle
' by finding the minimum and maximum x and y.
firstFlag = True
for each aPoint in pointList
aX = aPoint.GetX
aY = aPoint.GetY
if (firstFlag) then
minX = aX
maxX = aX
minY = aY
maxY = aY
firstFlag = False
else
minX = minX Min aX
maxX = maxX Max aX
minY = minY Min aY
maxY = maxY Max aY
end
end

'
' Make sure you can draw a circle through min/max
```

114 Programming View Documents

```
' X&Y.
if ((minX = maxX) OR (minY = maxY)) then
MsgBox.Error ("Cannot analyze a linear
distribution","CRM")
exit

end

'
' Get the center and determine a radius.
midX = ((maxX - minX)/2) + minX
midY = ((maxY - minY)/2) + minY
cenPoint = Point.Make (midX,midY)
radLen = (((maxX-midX)^(2)) + ((maxY-midY)^(2))).Sqrt
'
' Draw a circle.
aCircle = Circle.Make (cenPoint,radLen)
if (nil = aCircle) then
MsgBox.Error("Unable to create circle", "CRM")
exit
end

aGCircle = GraphicShape.Make (aCircle)
theGLIST.Add (aGCircle)
'
' Draw lines from center to each point
for each aPoint in pointList
aLine = Line.Make (cenPoint, aPoint)
aGLine = GraphicShape.Make (aLine)
theGList.Add (aGLine)
end
```

The *CRM_New_Analysis* script removes existing points so that new points can be entered. The script selects and deletes all graphical elements and removes all points from the list attached to the view.

```
' CRM_New_Analysis
'
```

Graphical Elements   115

```
' Get the display and graphics list of this view.
theView = Av.GetActiveDoc
theDpy = theView.GetDisplay
theGList = theView.GetGraphics
'
' Attach a new and empty list for points.
pointList = List.Make
theView.SetObjectTag (pointList)
'
' Delete all graphical objects.
theGlist.SelectAll
theGlist.ClearSelected
```

The Graphics List

You can draw point symbols, lines, text, and other graphical elements in a view document on top of any displayed theme. ArcView holds these elements in a list associated with the view's display. Avenue can access and manipulate the associated list. Once you have accessed the list, you can add or delete elements, or change the size, location, and orientation of elements. The following request accesses a view's graphical list.

```
aGList = theView.GetGraphics
```

Appearing below are samples of requests that can be made to a graphics list.

```
aGList.Add (aGraphicElement)
howMany = aGList.Count
aNewGList = aGList.GetSelected
aGList.SelectAll
aGList.UnselectAll
aGList.ClearSelected
```

Drawing Graphical Elements

Four steps are required to draw a graphical element on an existing view:

116   Programming View Documents

1. Retrieve the associated graphics list.

2. Create the object to draw, e.g., a line or circle.

3. Convert the object to a graphical element.

4. Add the element to the graphics list.

The code segment below shows how lines are drawn from a central point to various points stored in a list.

```
theGlist = theView.GetGraphics
' pointList holds objects of the Point class
' for each aPoint in the pointList.
aLine = Line.Make (cenPoint, aPoint)
aGLine = GraphicShape.Make (aLine)
theGList.Add (aGLine)
end
```

Text works in a similar fashion, as shown in the following code segment.

```
theGlist = theView.GetGraphics
aLabel = MsgBox.Input ("Point label:", "CRM", "")
aGLabel = GraphicText.Make (aLabel,aPoint)
theGList.Add (aGLabel)
```

Shapes that can be used to create graphical elements are listed below.

```
aCircle = Circle.Make (centerPoint, radiusLength)
aGCircle = GraphicShape.Make (aCircle)

aLine = Line.Make (startPoint, endPoint)
aGLine = GraphicShape.Make (aLine)

anOval = Oval.Make (originPoint, extentPoint)
aGOval = GraphicShape.Make (anOval)

aPoint = Point.Make (x, y)
aGPoint = GraphicShape.Make (aPoint)
```

Graphical Elements

```
aPolygon = Polygon.Make (aListOfVertices)
aGPolygon = GraphicShap.Make (aPolygon)

aRectangle = Rect.Make (originPoint, extentPoint)
aGRectangle = GraphicShape.Make (aRectangle)

aGText = GraphicText.Make (aTextString,
lowerLeftPoint)
aGText.SetBounds (aRectangle)
```

The **SetBounds** request defines the extent of a graphical element. When used with text, the request defines paragraph margins.

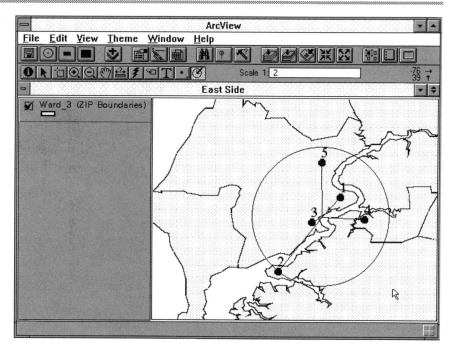

A simplified crime analysis application. (The coverage was provided by Geographic Data Technology, Inc.)

Chapter 7

Programming Themes

The access and manipulation of objects through Avenue scripts add powerful capabilities to any ArcView application. In this chapter we explore how to add themes, change theme properties, and select theme features.

Adding and Displaying Themes

Applications containing predetermined sets of themes are common. Your application should automatically add the theme and set its properties. This section demonstrates how you can add a theme to a view, duplicate it with different properties, and work with the theme legend. A script named *Load_Theme* is developed to demonstrate these features. Results of running the script are shown in the following figure.

120 Programming Themes

Adding and displaying themes.

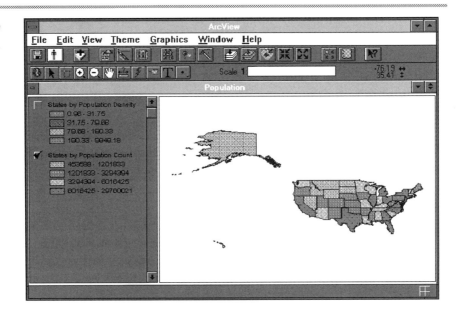

Script features are explained within the code segment below.

```
' Load_Theme
'
' This script is executed from a customized
' pushbutton on the view document.
'
theView = Av.GetActiveDoc
'
' Establish a data source and load it into the view.
aSrcName = SrcName.Make
("\avdata\namerica\usa\usa region.state")
aTheme = Theme.Make (aSrcName)
theView.AddTheme (aTheme)
'
' Duplicate the new themes, then
' change their names.
aTheme.SetActive (True)
```

Adding and Displaying Themes 121

```
theView.CopyThemes
theView.Paste
firstTheme = theView.GetThemes.Get(0)
secondTheme = theView.GetThemes.Get(1)
firstTheme.SetName ("States by Population Density")
secondTheme.SetName ("States by Population Count")
'
' Load the table with population data.
popVTab = VTab.Make
("\avdata\namerica\usa\stdemog.dbf".AsFileName,
False, False)
'
' Join the themes with the population table.
popJoinField = popVTab.FindField
("State_fips")
theDensityVTab = firstTheme.GetFTab
theCountVTab = secondTheme.GetFTab
for each oneVTab in {theDensityVTab,theCountVTab}
theJoinField = oneVTab.FindField
("State_fips")
oneVTab.Join (theJoinField, popVTab, popJoinField)
'
' Set each theme's legend;
' use quantile for the first one and
' use equal intervals for the second legend.
densField = theDensityVTab.FindField
("Pop90_sqmi")
popField = theCountVTab.FindField ("Pop1990")
firstLegend = firstTheme.GetLegend
firstLegend.Quantile (theDensityVTab, densField, 4)
secondLegend = secondTheme.GetLegend
secondLegend.Quantile (theCountVTab, popField, 4)
'
' Display the population count theme
' and make sure the legends are visible.
```

122   Programming Themes

```
firstTheme.SetVisible (False)
firstTheme.SetLegendVisible (True)
firstTheme.InvalidateLeggend
secondTheme.SetVisible (True)
secondTheme.SetLegendVisible (True)
secondTheme.InvalidateLegend
```

Data Source

Themes are created from data sources. ArcView supports a variety of data sources including raster image files, ArcStorm, or dynamic segmentation events. ArcView encapsulates these diverse data sources into objects of the **SrcName** class. A SrcName object identifies the data source used in creating a theme. The SrcName class has specialized subclasses to accommodate more complex data sources. ArcView data sources and identifying classes are listed below.

❏ Image file, SrcName

❏ Shape file, SrcName

❏ Coverage, SrcName

❏ Library layer, SrcName

❏ ArcStorm, SrcName

❏ X-Y event file, XYName

❏ Geocoding feature source, GeoName

❏ Dynamic segmentation source, DynName

A **SrcName** object has four attributes: file name, name, sub-name, and data source name. The file name attribute references the disk location of the data source, and is nil for a library layer. The name attribute is the name of the data source which becomes the name of the theme created from the SrcName. The sub-name attribute constitutes the feature class, i.e., **region.state**, **route.subclass**, **polygon**, **arc**, or **point**. Finally, the data source is the name of the coverage, file or library layer that contains the data. For coverages or files, the data source includes the full path to the coverage or file. In the case of a library layer, the data source has the following form: *LibraryName.LayerName*.

Adding and Displaying Themes   123

You can create a **SrcName** object with the **Make** request. The Make request requires a source string as its parameter, which varies for different data source types. The following code segment shows how to create a SrcName object for each source type.

```
' Image file.
aSrcName = SrcName.Make ("c:\images\myimage.gif")
'
' Shape file.
aSrcName = SrcName.Make ("c:\shapes\myshape.shp")
'
' Polygon coverage.
aSrcNamne = SrcName.Make ("c:\covrgs\usa polygon")
'
' ARC/INFO librarian.
aSrcName = SrcName.Make ("library.layer polygon")
'
' ArcStorm.
aSrcName = SrcName.Make ("librarian.library.layer
polygon")
'
' A region.
aSrcName = SrcName.Make ("c:\avdata\usa
region.state")
```

Once a **SrcName** object is created you can add it to a view as a theme. The following statements show how to add a theme from the SrcName object.

```
aTheme = Theme.Make (aSrcName)
aView.AddTheme (aTheme)
```

The **SrcName** class also provides requests that retrieve information about an existing source name. If you are uncertain about the required properties, you can manually load your theme. Use the following Avenue script to learn more about the properties of the theme's source name.

124   Programming Themes

```
' Get_SourceName_Info
'
' This script is executed from a customized
' pushbutton in a view document. It assumes
' that your theme is the first one in the
' view's table of contents.
theView = Av.GetActiveDoc
theTheme = theView.GetThemes.Get(0)
theSrcName = theTheme.GetSrcName
'
' Get source name's properties.
theName = theSrcName.GetName
theFile = theSrcName.GetFileName
theDataSource = theSrcName.GetDataSource
theSubName = theSrcName.GetSubName
'
' Place these properties in a list to display.
aList = List.Make
aList.Add ("Name:"++theName)
aList.Add ("File:"++theFile.GetFullName)
aList.Add ("Source:"++theDataSource)
aList.Add ("Sub:"++theSubName)
MsgBox.ListAsString
(aList,"Source name properties:","")
```

Legends

There are two ways to establish legends. First, you can manually create a legend in ArcView and store it as a legend file for later retrieval and use in Avenue. Second, you can access a theme's legend and modify it through object requests.

The following code shows how to load a legend file. This approach is useful in cases where the legend does not change every time the application is used, or where a basic legend is the starting point for creating a more complex and dynamic version.

Adding and Displaying Themes 125

```
' Load_Legend
'
' This script is executed from a customized
' pushbutton in a view document and assumes
' that the legend applies to the first theme
' on the view's table of contents.
theView = Av.GetActiveDoc
theTheme = theView.GetThemes.Get(0)
theLegend = theTheme.GetLegend
aLegendFile = "myleg.avl".AsFileName
theLegend.Load (aLegendFile)
```

Once you have the legend object to work with, you can set its properties through legend object requests. These requests are explained in the following code segments.

```
' Return the legend to its original form.
aLegend.Default
```

```
' Use unique classes for every value using
' a field of an attribute table.
aLegend.SetMaxClassifications (10)
didIt = aLegend.Unique (anFTab, aField)
' didIt is True if request was successful,
' and number of unique values was limited to 10.
```

```
' Use an equal number of values per classification.
aLegend.Quantile (anFTab, aField, numClasses)
```

```
' Use equal interval classifications.
aLegend.Interval (anFTab, aNumericField, numClasses)
```

Copying

A theme is a pointer to a data source; therefore, copying a theme does not copy the data but simply generates new pointers. Theme copying provides a powerful feature to view diverse aspects of a single data source.

126   Programming Themes

Themes can be independently queried to show a particular aspect of the data source. For example, a point coverage that contains insurance company agents and the years their offices were opened can hold several themes displaying offices opened in a specific year or years.

CopyThemes and ***Paste*** requests are sent to a view object to copy the active themes to a clipboard and then paste them to the view at the beginning of the view's table of contents. In the following code segment all themes are copied from one view to another.

```
' Make all themes active.
themesList = theView.GetThemes
for each aTheme in themesList
aTheme.SetActive (True)
end

theView.Copythemes
anotherView.Paste
```

The ***CutThemes*** request removes the active themes from the view, and places them in the clipboard.

Using Queries

Avenue can set and evaluate a query expression for a theme. Applications requiring dynamically changing queries can use this feature. In the following script, Avenue sets a query that would select all states with equal or greater population than a pre-selected state.

```
' Set_Query
'
' This script is executed from a pushbutton on
' the view's user interface. A state
' must be selected beforehand.
'
' Get the basic information.
theView = Av.GetActiveDoc
```

```
theTheme = theView.FindTheme ("State")
theDpy = theView.GetDisplay
theVTab = theTheme.GetFTab
theBitMap = theVTab.GetSelection
'
' Retrieve the population of the
' selected state.
if (theBitMap.Count <> 1) then
MsgBox.Warning ("Select one state","Set_Query")
exit
end

selState = theBitMap.GetNextSet (-1)
popField = theVTab.FindField ("Pop1990")
selPop = theVtab.ReturnValueString (popField,
selState)
'
' Set a query for states of equal or
' greater population. Place field names
' in square brackets [ ].
aQStr = "[Pop1990] >="++selPop
' Apply the query to the theme.
theVTab.Query (aQStr, theBitMap, #VTAB_SELTYPE_NEW)
theVTab.SetSelection (theBitMap)
```

Selecting Features

ArcView's default user interface for view documents contains a tool button for selecting features. In Avenue, features can be selected by accepting a point, a line, a rectangle (box), or a polygon from the user.

128 Programming Themes

Point Selection

SelectByPoint is the request sent to a theme to select a spatial feature. In the following code segment, a feature is selected from a location identified by pointing the mouse. The feature is added to the list of selections.

```
' Point_Selection
'
' This script is executed from a tool button
' of a view document. It selects features from
' the first theme on the table of contents.
theView = Av.GetActiveDoc
'
' The view's display object is required to get a
' mouse location.
theDpy = theView.GetDisplay
'
' Get the first theme.
theTheme = theView.GetThemes.Get(0)
'
' Make the selection.
theTheme.SelectByPoint
(theDpy.GetMouseLoc, #VTAB_SELTYPE_OR)
```

The second parameter in the *SelectByPoint* request determines whether the selection set is new or added to current selections. Valid values for this parameter follow:

```
' Create a new selection set.
#VTAB_SELTYPE_NEW

' Select from the current selection set
#VTAB_SELTYPE_AND

' Add to the current selection set.
#VTAB_SELTYPE_OR
```

```
' If the selected feature is not currently
' selected, then add it to the current
' selection set. Otherwise, remove it from
' the current selection set.
#VTAB_SELTYPE_XOR
```

The first parameter for the **SelectByPoint** request is a point object. The **GetMouseLoc** returns a point which is the location of a mouse click.

Line Selection

Selecting features with a line is similar to the point selection. As shown in the statement below, line selection differs from point selection in that the **SelectByLine** request applies only to an object of the feature table.

```
theTheme.GetFTab.SelectByLine
(theDpy.ReturnUserLine, #VTAB_SELTYPE_NEW)
```

Box Selection

The **SelectByRect** request selects features inside a rectangle or touched by the rectangle's sides. You can supply the rectangle by constructing it, or the user can select it. In the following script, the user selects a rectangular area.

```
' Box_Selection
'
' This script is executed from a tool button
' of a view document. It selects features from
' the first theme on the table of contents.
theView = Av.GetActiveDoc
'
' Get the first theme.
theTheme = theView.GetThemes.Get(0)
'
' Ask user for a rectangle.
aBox = theView.ReturnUserRect
```

130 Programming Themes

```
'
' Make the selection.
theTheme.SelectByRect (aBox, #VTAB_SELTYPE_NEW)
```

The ***ReturnUserRect*** request pauses the script and transfers control to the view document. The user then clicks two points on the view's display to establish a rectangle. This rectangle is then returned to the script.

Polygon Selection

Selecting features with a polygon is similar to the box selection. The difference is in the request. The following statement selects features using a polygon chosen by the user.

```
theTheme.SelectByPolygon (theView.ReturnUserPoly,
#VTAB_SELTYPE_NEW)
```

Chapter 8

Programming Table Documents

Because ArcView maintains attribute data in **Table** documents, the access and manipulation of attribute data are implemented through these documents. In this chapter, we will develop an application that assigns the polygons of a coverage to territories or districts. This type of application is useful for redistricting, assigning sales territories, and similar projects. In the process, you will learn how to create tables, retrieve or set attribute values, and join tables.

Assigning Polygons to Territories

Avenue scripts in this chapter use the *usa* coverage bundled with the ArcView software. The coverage is located in ArcView's *avdata\namerica\usa* directory.

132 Programming Table Documents

The scripts described below are developed in the code segments to follow. Script results are shown in the next figure.

Create_Att_Table creates a new attribute table with two fields, one for the join operation and the other to hold a territory number. Because this script is used only once for each new coverage, execution from a menu item is more appropriate than from a pushbutton.

Load_Att_Table imports the newly created attribute table and joins it with the theme's table. This script is also executed from a menu item since it is needed only once for each new coverage.

Assign_Territory sets the territory field number of the selected polygons to a user-provided value. The user can utilize any of the selection tools such as point, circle, or polygon to select appropriate polygons. The script deselects all polygons after setting the territory number.

```
' Create_Att_Table
'
' This script creates a new table and copies
' values of a field to be joined later.
'
' Get a file name to create.
aFileName = FileDialog.Put("terr.dbf".AsFileName,
"*.dbf","Create Attribute Table")
if (nil = aFileName) then
exit
end

'
' Create a dBase table.
attVTab = VTab.MakeNew (aFileName,dBase)
attVTab.SetEditable (True)
fipsField = Field.Make
("State_fips", #FIELD_CHAR, 2, 0)
terrField = Field.Make
("Terr_num", #FIELD_CHAR, 4, 0)
attVTab.AddFields ({fipsField, terrField})
```

Assigning Polygons to Territories   133

```
'
' Get State_fips values from the theme's table
' and write to the new table.
thisView = Av.GetActiveDoc
theTheme = thisView.GetThemes.Get(0)
theVTab = theTheme.GetFTab
stateField = theVTab.FindField("State_fips")
if (nil = stateField) then
MsgBox.Error ("Unable to find the state field",
"")
exit
end

' The record count is required for a loop.
recordCount = theVTab.GetNumRecords
if (0 = recordCount) then
MsgBox.Error ("There are no records to copy",
"")
exit
end

' Read and write each record
for each index in theVTab
fieldValue = theVtab.ReturnValueString
(stateField,index)
if (nil = fieldValue) then
continue
end

' Add territory number zero to each
' record as default.
rec = attVTab.AddRecord
attVTab.SetValue (fipsField, rec, fieldValue)
attVTab.SetValue (terrField, rec, "0")
end
```

134 Programming Table Documents

```
attVTab.Flush
'_____-

' Load_Att_Table
'
' This script imports the attribute table and
' joins it with the theme's table.
'
' Get the theme's table.
thisView = Av.GetActiveDoc
theTheme = thisView.GetThemes.Get(0)
theVTab = theTheme.GetFtab
theJoinField = theVtab.FindField("State_fips")
if (nil = theJoinField) then
MsgBox.Error ("Theme's table does not have the join
field",
"")
exit
end

'
' Load the attribute table.
attTableFile = FileDialog.Show
("*.dbf","dBase Files",
"Load Attribute Table")
if (nil = attTableFile) then
exit
end

'
' Make sure a table can be created
' from the given file name.
isOK = VTab.CanMake (attTableFile)
if (isOK.Not) then
MsgBox.Error ("Invalid file","")
```

Assigning Polygons to Territories 135

```
exit
end
'
' Load the table.
forWrite = True
skipFirst = False
attVTab = VTab.Make (attTableFile, forWrite,
skipFirst)
attTable = Table.Make (attVTab)
if (nil = attTable) then
MsgBox.Error ("Unable to load attribute table",
"")
exit
end

attTable.SetName ("Territory")
attJoinField = attVTab.FindField ("State_fips")
if (nil = attJoinField) then
MsgBox.Error ("Attribute table does not have the
join field",
"")
exit
end
'
' Join the tables and display the result.
theVTab.Join (theJoinField, attVTab, attJoinField)
theTheme.EditTable
'_____-

' Assign_Territory
'
' This script will set the selected records
' to a specified territory number.
thisView = Av.GetActiveDoc
theTheme = thisView.GetThemes.Get(0)
```

136 Programming Table Documents

```
theVTab = theTheme.GetFTab
theVTab.SetEditable (True)
isOK = theVTab.IsEditable
if (isOK.Not) then
MsgBox.Error ("Cannot edit the table", "")
exit
end

' Make sure at least one feature is selected.
theBitMap = theVTab.GetSelection
selectedCount = theBitMap.Count
if (0 = selectedCount) then
MsgBox.Warning ("Select areas first", "")
exit
end
'
' Get a territory number.
terrNum = MsgBox.Input ("Enter a territory number: ",
"","0")
if (nil = terrNum) then
exit
end
'
' Get the field to edit.
attVTab = av.GetProject.FindDoc("Territory").GetVTab
attVTab.SetEditable (True)
theField = theVTab.FindField("Terr_num")
theField.SetEditable(True)
'
' Set the value of the selected records
for each index in theVTab.GetSelection
attVTab.SetValueString(theField, index, terrNum)
end
'
```

```
' Clear the selection.
theBitMap.ClearAll
theVtab.SetSelection (theBitMap)
'_____
```

Assigning polygons to a territory.

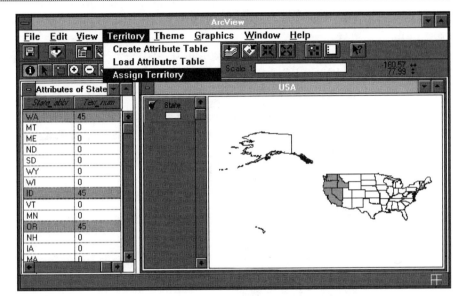

Table Document

Tables provide the means to manipulate the data associated with spatial elements. You will need to access such data in most of your applications. Tables are document objects that serve as an interface to the underlying database.

Upon opening a **Table** document, a ***VTab*** is displayed in a format resembling a spreadsheet. VTab objects are virtual tables composed of tabular data stored on disk outside ArcView.

Creating a Table

You can create a **Table** document from existing disk files or from new disk files. If you are creating a new table in a new disk file, your choices are comma delimited text, dBase, or INFO files. In the case of creating Table documents from existing files, your choices include an SQL table, ARC/INFO Librarian, ArcStorm, dBase files, INFO files, and delimited text files.

To create a new delimited text file, use the **_FileDialog_** class to establish a file name. Create the file and write the field names on the first line as shown in the following code segment.

```
' Get a file name to create.
aFileName = FileDialog.Put("terr.txt".AsFileName,
"*.txt","Create Attribute Table")
if (nil = aFileName) then
exit
end

'
' Create a delimited text file.
aLineFile = LineFile.Make(aFileName,
#FILE_PERM_WRITE)
'
' Write field names as the first line.
aLineFile.WriteElt("State_abbr,Terr_num")
```

You can continue writing data into the new file. However, once the file is created it can be treated as an existing file. The code segment below creates a new table document by loading a disk file into the ArcView project.

```
newVTab = VTAB.Make (aFileName, forWrite, skipFirst)
newTable = Table.Make (newVTab)
```

Displaying a Table

Users can open a **Table** document manually from the project window. However, an existing table is not necessarily listed in the project window.

Table Document   139

Therefore, if you want to give users the option of manually opening the table, you should ask Avenue to add the document to the project window. The following code line is an example of how to add a table to the project window.

```
Av.GetProject.AddDoc (newTable)
```

The process of opening or closing a **Table** document can also be automated through Avenue. If the table is associated with a theme, you can use the *EditTable* request to open the theme's table as shown below.

```
aTheme.EditTable
```

A more direct method to control document display is through its window. As demonstrated in the following code segment, once you have a table's *DocWin* object, you can then open, close, resize, and move the table, as well as carry out other operations.

```
' attTable is a table document.
' Get its document window object.
attDocWin = attTable.GetWin
'
' Open the object if it is closed.
if ( attDocWin.IsOpen.Not ) then
attDocWin.Open
end
'
' Minimize the table.
attDocWin.Minimize
'
' Restore the table from its icon form.
attDocWin.Restore
'
' Make the table the active window
attDocWin.Activate
'
' Move the document around.
```

140   **Programming Table Documents**

```
attDocWin.MoveTo ( newX, newY)
'
' Close the document.
attDocWin.Close
```

Setting Table Properties

Table properties include name, creator, comments, and object tag. Similar to other document properties, table properties are generally used for documentation. At the very least, you should set the name property because name is always visible. Setting the creator and comments properties are recommended.

The object tag property can be useful for sharing data between Avenue scripts. For instance, you may require the user to run the *Clean_Up* script before the *Compute* script. One way to enforce this sequence is to tag the CLEAN string to the table after running *Clean_Up*, and then check for the string when *Compute* is executed.

The following code lines show the Avenue requests for setting table properties.

```
' newTable is a Table object.
' Set its properties.
'
newTable.SetName ("District Detail")
newTable.SetCreator ("A. H. Razavi")
newTable.SetComments
("Store characteristics of each district in this
table")
'
' Set the object tag to NEW string.
newTable.SetObjectTag ("NEW")
```

As demonstrated below, you can access the properties of an existing table through **Get** requests.

```
' Table name is a string object.
aName = aTable.GetName
whoCreated = aTable.GetCreator
```

```
whenCreated = aTable.GetCreationDate
tableComments = aTable.GetComments
someObject = aTable.GetObjectTag
```

Joining Tables

Spatial analyses often require information from more than one table. Multiple tables that are to be used simultaneously must be joined. The most frequent join operation involves a theme's table and a non-spatial attribute data file. Avenue can automate the join process for your user. In the following code segment, the join fields from two tables are identified, and then the two tables are joined through their respective **VTabs**.

```
' Get the theme's table information.
thisView = Av.GetActiveDoc
theTheme = thisView.GetThemes.Get(0)
theVTab = theTheme.GetFtab
theJoinField = theVtab.FindField("State_fips")
'
' Get the attribute table information.
attVTab = attTable.GetVTab
attJoinField = attVTab.FindField ("State_fips")
'
' Join the tables and display the result.
theVTab.Join (theJoinField, attVTab, attJoinField)
```

As presented in the following code line, the **Join** request for a **VTab** requires three parameters. The parameters are described below.

```
aToVtab.Join (aToField, aFromVtab, aFromField)
```

aToField is the common field from the destination table which receives the result of the join operation (*theJoinField* in the preceding example).

aFromVTab is the source table for the join operation (*attVTab* above).

142 Programming Table Documents

aFromField is the common field from the source table (*attJoinField* in the example).

To join more than two tables, join two, and then join the third table to the result and so on. A join operation assumes that the relation between the two tables is one-to-one. However, the relation may also be many-to-one where each record of the source table matches one or more records of the destination table.

If the relation is many-to-one, i.e., when one or more matching records in the source table exist(s) for each record, you must use the link instead of the join operation. A link operation may also be applied when the relation is one-to-one or many-to-one. The link operation matches selected records, while join operates on the entire table. Next, link does not provide a virtual table comprised of the source and destination tables. The link operation, however, does return the linked information for the **Select** or **Identification** operations.

Sorting Records

Sorting a table can improve readability, and may be required in your application. An example would be displaying states in ascending order of population size. The ***Sort*** request accepts the sort field and Boolean parameters. If the Boolean parameter is True, the sort is arranged in descending order; if False, the sort will be arranged in ascending order.

```
myTable.Sort (sortField, isItDescending)
```

Printing Tables

Generating reports requires the ability to query the table, select fields, and format the results. ArcView allows you to apply queries to tables, and through Avenue you can output selected fields. Formatting the report, however, is not possible. You should consider exporting the entire table or selected records to an external file so that a third party report writer could be used to generate reports.

Printing Tables 143

To print a table into an external file, use the ***Export*** request as demonstrated in the following script.

```
' Export_Table
'
' This script exports the themes table.
thisView = Av.GetActiveDoc
theTheme = thisView.GetThemes.Get(0)
theVTab = theTheme.GetFTab
if (nil = theVTab) then
MsgBox.Error("Unable to open theme's table",
"")
exit
end

'
' Get a file name.
aFileName = FileDialog.Put( "terr.dbf".AsFileName,
"*.dbf","Create External Table")
if (nil = aFileName) then
exit
end

'
' Export the table to a dBase file.
theVTab.Export (aFileName, dBase, False)
```

As seen in the following code lines, the ***Export*** request requires three parameters. The parameters are described below.

```
aVTab.Export (aFileName, aTableClass,
selectedRecordsOnly)
```

- ❏ *aFileName* is the name of the destination file for the export operation (*aFileName* in the preceding example).
- ❏ *aTableClass* can be a dBase, INFO, or DText value (*dBase* above).

144   Programming Table Documents

❏ *selectedRecordsOnly* is a Boolean object that determines if the entire table is exported or only selected records (*False* in the preceding example). A False value exports the entire table.

Chapter 9

Accessing Databases

Tabular databases outside ArcView can be viewed and sometimes edited through table documents. ArcView tables do not hold the actual data but rather manage a view of a tabular data source. Tables are also dynamic, meaning that changes to the data source are reflected in the ArcView table. In addition, edits to the table document directly affect the source.

Chapter 8 reviewed the importance of attribute data and how to use such data in your applications. In this chapter, we examine how to access data from a variety of sources. ArcView can use data sources such as dBase, INFO, and delimited text files. ArcView can also connect to SQL databases and run SQL queries. Supported SQL databases include AS400, Informix, Ingres, Oracle, and Sybase. The last section of this chapter shows how to access ArcStorm, the ARC/INFO database.

146   Accessing Databases

Creating File Based Database Objects

The purpose of a table document in ArcView is to display a ***VTab***. The VTab object, also known as a "virtual table", manages the view of the tabular data source. VTabs may be joined or linked to form new views of the data. These operations were discussed in Chapter 8.

To create a database object, you must first create a VTab representing the data. A VTab object can be created based on an existing data source, or a new data source can be simultaneously created. The Avenue script below shows how to create a VTab object from an existing data source or sources.

```
' Ask user to select existing
' data source(s) from
' dBase, or delimited text. Add
' each data source to the project.
fileLabels = {"dBase","Delimited Text", "All Files")
filePatterns = {"*.dbf","*.txt", "*.*"}
fileNameList = FileDialog.ReturnFiles
(filePatterns, fileLabels,
"Select Data Sources", 0)
for each fileName in fileNameList
if ( VTab.CanMake (fileName) ) then
' Create a VTab in read only mode.
aVTab = VTab.Make (fileName, False, False)
if (aVTab.HasError) then
MsgBox.Error
("Unable to create a database object from"++
fileName.AsString, "")
else
' Create a table from the VTab.
aTable = Table.Make (aVTab)
aTable.SetName (fileName.GetBaseName)
end
```

Creating File Based Database Objects 147

```
end
end
```

In the preceding script, the ***Make*** request to the ***VTab*** class requires the three parameters listed below.

```
myVTab = VTab.Make (aFileName,
forWrite,
skipFirstRecord)
```

If set to True, the second parameter creates the database as an editable object. The value of this parameter is returned by the ***CanEdit*** request. However, setting this parameter does not guarantee that a VTab can be edited. The parameter acts a security measure to allow editing. You must first ask to start editing with the ***SetEditable*** request. The ***IsEditable*** request returns True if indeed the VTab object can be edited. These requests are presented below.

```
' allowedToEdit and canModify are
' Boolean objects.
allowedToEdit = myVTab.CanEdit
myVTab.SetEditable (True)
canModify = myVTab.IsEditable
```

If set to True, the third parameter in the ***Make*** request causes VTab to skip the first record. This feature is often useful with text files that hold field names in the first line of the file.

Two other new requests, ***CanMake*** and ***HasError***, can also be used in the preceding script. The CanMake request to the VTab class returns True if a VTab can be created from a file name; otherwise, it returns False. The HasError request to a VTab object returns True if there is a problem in creating the VTab object; otherwise, it returns False.

A new data source file can be created by using Avenue. New files are often created to export existing data. The following Avenue script exports selected records in a table document.

```
' Ask user to select an export format
' from INFO, dBase, or Delimited Text;
' retrieve a new file name; create table;
```

148 Accessing Databases

```
' and export the selected records.
theTable = av.GetActiveDoc
fileFormat = MsgBox.ChoiceAsString
({"INFO","dBase","Delimited Text"},
"Select an export format:", "")
if (fileFormat = "INFO") then
fileClass = INFO
filePattern = "arcdr9"
fileExtension = ""
elseif (fileformat = "dBase") then
fileClass = dBase
filePattern = "*.dbf"
fileExtension = "dbf"
elseif (fileformat = "Delimited Text") then
fileClass = DText
filePattern = "*.txt"
fileExtension = "txt"
else ' User clicked on Cancel button.
exit
end

' Get a new file name based on the selected format.
newFileName = FileDialog.Put
(("new."+fileExtension).AsFileName,
filePattern, "Export to a New File")
if (nil = newFileName) then
' User clicked on Cancel button.
exit
end

theVTab = theTable.GetVTab
theVTab.Export (newFileName, fileClass, True)
```

The **_Export_** request to a VTab object requires three parameters. The first parameter is a new file name. The second determines the type of file, i.e., INFO, dBase, or delimited text. The third parameter is a Boolean object. If the Boolean object is set to True, only the selected records are

Creating File Based Database Objects 149

exported; if set to False, all records are exported. The structure of the new database file is the same as the VTab's structure.

You can create a new database file and then directly build its structure by adding fields. In the following script, a dBase file is created to maintain county populations.

```
' The new dBase file has four fields.
' Create the field objects, create the dBase
' file, and then add the fields to the file.
field1 = Field.Make
("COUNTY", #FIELD_CHAR, 20, 0)
' Total population field.
field2 = Field.Make
("POP_TOTAL", #FIELD_LONG, 8, 0)
' Population age 18 years and over.
field3 = Field.Make
("POP_18", #FIELD_LONG, 8, 0)
' Percent of 18+ population.
field4 = Field.Make
("PCT_18", #FIELD_DECIMAL, 6, 2)
' Move all fields into a field list.
fieldList = {field1, field2, field3, field4}
' Create a new data source file and object.
newVTab = VTab.MakeNew ("pop.dbf",dBase)
if (newVTab.HasError) then
exit
end

' Add the fields to the new dBase file.
if (newVTab.CanAddFields) then
newVTab. AddFields (fieldList)
end
```

The **Make** request to the **Field** class requires four parameters. The first and second parameters are the field name and the field type enumeration. The third and fourth parameters are the "width" and

150   Accessing Databases

precision of the field. For example, in the preceding script, *PCT_18* is a decimal field of six digits with two decimal places.

The ***MakeNew*** request to the VTab class requires two parameters: a file name, and a file class. File classes include ***INFO***, ***dBase*** and ***DText***.

Reading Records and Fields

Once a VTab object is created from a tabular data source, the VTab can access the entire database. A table document displays the records accessed by the VTab. The VTab's view of data records is limited only if a definition query is associated with the VTab.

In this section, the discussion includes how to access field values for use inside your application. In the following code segment, the difference between states with the highest and lowest populations is computed.

```
' To retrieve a field value, the
' field object is required.
theTable = av.GetProject.FindDoc("stdemog.dbf")
theVTab = theTable.GetVTab
stateField = theVTab.FindField ("State_name")
popField = theVTab.FindField ("Pop1990")
' Get the highest and lowest values by sorting.
' A descending sort places the highest value in
' row 1.
theTable.Sort (popField, True)
recordNumber = theTable.ConvertRowToRecord (0)
mostState = theVtab.ReturnValueString
(stateField, recordNumber)
mostPopulation = theVTab.ReturnValueNumber
(popField, recordNumber)
' The lowest value is in the last row.
recordNumber = theTable.ConvertRowToRecord
((theVTab.GetNumRecords)-1)
```

Reading Records and Fields 151

```
leastState = theVtab.ReturnValueString
(stateField, recordNumber)
leastPopulation = theVTab.ReturnValueNumber
(popField, recordNumber)
' Report the difference.
MsgBox.Info (mostState++"has"++
(mostPopulation-leastPopulation).AsString++
"more people than"++leastState, "")
```

When a table sorts or promotes **VTab** records, record order does not physically change. Instead, changes in record order are displayed. Thus, there is a difference between table document rows and VTab records. In the preceding code segment, the table was sorted in such a way that the record with the highest population was placed in the first row. However, the sort did not change the record number in VTab. The **ConvertRowToRecord** request was used for this purpose. When the row of the least populated state was converted to record number, **GetNumRecords-1** was used as the number for the last row in the table. The GetNumRecords request to a VTab returns the number of records.

The **ReturnValue** request returns an object matching the class of the field value. You can force the return value to be of the **Number** or **String** class by using **ReturnValueNumber** or **ReturnValueString** requests against a **VTab**. Both requests accept two parameters: a field object and a record number.

The **Identify** window is used to display a specific record. In the following code segment, the user selects a row from a table, and the row is displayed in the Identify window.

```
rowNumber = theTable.GetUserRow
recordNumber = theTable.ConvertRowToRecord
(rowNumber)
theVTab.Identify (recordNumber, "")
Editing Records
```

You can change the value of a field or add new records to the VTab. When the VTab is edited, the physical data source is modified. Before editing a VTab, always verify whether it can be edited by using the following requests:

152 Accessing Databases

```
' These requests return a Boolean object.
allowedToEdit = myVTab.CanEdit
canModify = myVTab.IsEditable
canAddToIt = myVTab.CanAddRecord
```

When setting a field value, you will generally wish to set it for all records or a certain group of records. You can step through all or a selected group of records, and set a field value for each. The following statements show how to step through records.

```
' Step through selected records.
for each offset in theVTab.GetSelection
theVTab.SetValue (aField, offset, valueObject)
end

'
' Step through all records.
for each recordNumber in theVTab
theVTab.SetValue (aField, recordNumber,
valueObject)
end
```

If you are computing a field value for all or a group of records, it is easier to use the **Calculate** request. The Calculate request accepts a calculation string and a field to hold the results. The calculation string is the same expression that you could enter in ArcView's calculator. In the following code segment, the proportion of the population over age 18 is computed. The Calculate request applies to the selected records; if no records are selected, the request applies to all records.

```
' Calculate the proportion of the
' population over 18 for all records.
theVTab.GetSelection.ClearAll
theVTab.Calculate
("100*([POP_TOTAL]-[POP_18])/[POP_TOTAL]",
theVTab.FindField("PCT_18") )
```

To add a record, use the **AddRecord** request. The request adds a null record. Values can then be set for each field as shown in the following code segment.

```
newRecordNumber = theVTab.AddRecord
theVTab.SetValueString
(countyField, newRecordNumber, "FAIRFAX")
```

The **SetValue** request's third parameter requires an object that matches the field type. In order for the new value to be a **Number** or **String** class, use the **SetValueNumber** or **SetValueString** request to a **VTab**. Both requests accept three parameters: a field object, a record number, and the new value.

Accessing an SQL Database

The access procedure for SQL databases is slightly different from the procedure for file based data sources. However, once you have created the **VTab** object for an SQL database, the VTab can be used as discussed earlier in this chapter. The following steps describe how to connect to an SQL database and create a VTab object.

1. Verify that the SQL is available by executing the Avenue statement shown below:

```
isThereSQL = SQLCon.HasSQL
```

An **SQLCon** object manages the connection to an SQL database. The **HasSQL** request to the **SQLCon** class returns a Boolean object. The Boolean object is set to True if the SQL feature is available in your ArcView version; otherwise, its value is False.

2. Connect to a database by creating an **SQLCon** object. A list of databases available for connection should have been placed in the *default.db* file by your system manager. Avenue searches for this file to obtain the list of available databases and their parameters. You can either get the list or search for a specific one as shown in the following code segment.

154 Accessing Databases

```
' If Oracle server is not available, then
' display a list of available databases to
' the user for selection.
' In Windows, you must run this script from
' a menu item or pushbutton.
mySQLConnection = SQLCon.Find ("oracle")
if (mySQLConnection.HasError) then
' Connection failed.
listSQLCon = SQLCon.GetConnections
if (listSQLCon.Count = 0) then
MsgBox.Error
("Unable to find any SQL connection",
"")
exit
else
mySQLConnection = MsgBox.Choice (listSQLCon,
"Select a database:", "")
if (nil = mySQLConnection) then
exit
end
if (mySQLConnection.HasError) then
MsgBox.Error ("Unable to make the connection",
"")
exit
end
end
end
```

The **Find** request accepts a string parameter as the database name. The database name must match the name appearing in the *default.db* file. If you are not certain about the database name, send a **GetConnections** request for a list of all available database connections.

3. Log into the server by sending a **Login** request to your **SQLCon** object. The Login request needs a parameter as the login string. The

code line below includes the login to the Oracle database with the user name "system", and the password "manager".

```
mySQLConnection.Login ("system/manager")
```

4. Create a **VTab** object based on an SQL query. In the following code segment, a VTab is created for properties valued at over $50,000.

```
anSQLQuery = "select * from property"++
"where land_value > 50000.00"
mySQLVTab = VTab.MakeSQL (mySQLConnection,
anSQLQuery)
```

ArcView stores the definition of your SQL query and automatically reconnects to the database each time you open the project.

Using ArcStorm

ArcStorm (Arc Storage Manager), ARC/INFO's newest spatial database manager, was introduced with ARC/INFO version 7. Avenue's **Librarian** class is the equivalent of an ArcStorm database in ARC/INFO. Librarians contain zero or more libraries, libraries hold zero or more layers, and layers hold zero or more data sources. Each data source has a feature class such as line, polygon, or point. The data source can become an ArcView theme.

For example, the state of Virginia could be an ArcStorm database, and each county a library in the database. The layers in each library could be census tracts, census blocks, zip codes, major highways, and streams. The tracts and blocks layers would have source data for line and polygon feature classes. The zip codes layer would have a point feature class as the centroid of each area, in addition to line and polygon features. Finally, the line feature class would be the only data source for the highways and streams layers.

Use the following steps to access a data source from an ArcStorm database:

156   Accessing Databases

1. Select the ArcStorm database by creating a **_Librarian_** object. In the code segment below, the user is presented with a list of librarians from which to select.

```
' This request returns a list of accessible
' librarians in the ARCHOME\arcstorm directory.
listLibrarians = Librarian.ReturnLibrarians
if (listLibrarians.Count = 0) then
MsgBox.Error
("Unable to find any ArcStorm databases",
"")
exit
end

myLibrarian = MsgBox.Choice (listLibrarians,
"Select an ArcStorm database:", "")
```

If you know the database name, you can create the Librarian object by using the **_Make_** request. The code line below creates a Librarian object for the Virginia ArcStorm database.

```
' ARCHOME\arcstorm directory is assumed
' if path is not provided.
myLibrarian = Librarian.Make ("virginia")
```

2. Select the library by creating a **_Library_** object. In the code segment below, the user is presented with a list of libraries from which to select.

```
listLibraries = myLibrarian.ReturnLibraries
if (listLibraries.Count = 0) then
MsgBox.Error ("Unable to find any libraries in"++
myLibrarian.GetName++"database",
"")
exit
end

myLibrary = MsgBox.Choice (listLibraries,
"Select a library:", "Developer's Guide")
```

Using ArcStorm   **157**

If the library name is known, you can create the **Library** object by using the **Make** request. In the following code segment, a Library object is created for the Fairfax County library.

```
' Verify whether Fairfax library exists.
foundIt = Library.Exists ("fairfax")
if (Not foundIt) then
myLibrary = Library.Make ("fairfax", myLibrarian)
end
```

3. To select a layer in the library, create a **Layer** object. In the code segment below, the user is presented with a list of layers from which to select.

```
listLayers = myLibrary.ReturnLayers
if (listLayers.Count = 0) then
MsgBox.Error ("Unable to find any layers in"++
myLibrary.GetName++"library",
"")
exit
end

myLayer = MsgBox.Choice (listLayers,
"Select a layer:", "")
```

4. A layer can have several data sources, or one data source for each feature class. Access the data source by creating a **SrcName** object. In the following code segment, the user is requested to select a feature class to add to a view document.

```
listSrcNames = Layer.ReturnSrcNames
(myLayer.GetFullName)
if (listSrcNames.Count = 0) then
MsgBox.Error ("Unable to find any source in"++
myLayer.GetFullName, "")
exit
end
```

158   Accessing Databases

```
listFeature = List.Make
for each source in listSrcNames
' Prepare a list of feature classes.
listFeature.Add (source.GetSubName)
end
aFeature = MsgBox.ChoiceAsString
(listFeature, "Select a feature class:", "")
if (nil = aFeature) then
exit
end
mySrcName = SrcName.Make
(myLayer.GetFullName++aFeature)
```

The **GetFullName** request to a **Layer** object returns a string in the following format: *LibrarianName.LibraryName.LayerName*. The **Get-SubName** request to a **SrcName** object returns the feature class name. The **Make** request for a SrcName object needs one string parameter; the string parameter depends on the source type such as ArcStorm, coverage or shape file. Formats for the string parameter by source appear below.

```
Source: ArcStorm
Parameter: LibrarianName.LibraryName.LayerName
FeatureClass

Source: coverage
Parameter: FullPath\CoverageName FeatureClass

Source: shape file
Parameter: FullPath\ShapeFileName
```

Adding the **SrcName** object to a view document and accessing the object's attribute data are easy, as shown in the following code segment:

```
newTheme = Theme.Make (mySrcName)
aView.AddTheme (newTheme)
newVTab = newTheme.GetFTab
```

The **GetFTab** request to a feature theme returns a **VTab** object as the attribute data.

Programming Chart Documents

Charts provide a powerful visual presentation of attribute data. This chapter covers the creation, access and manipulation of chart documents through an application exercise that creates two charts of Maryland counties. One of the charts compares the number of students enrolled in the public and private school systems, and the other shows expenditure per student.

Scripts for Creating Charts

Appearing below is the entire Avenue script to create the two charts. Once developed, the script can be executed from the project window or through an interface control item such as a pushbutton.

160　Programming Chart Documents

```
' Education charts
'
' This script creates two charts: public versus
' private school enrollment and public expenditure
' per student. Both charts along with the view
' document are opened and arranged within the
' ArcView window. The user can then
' select counties and examine related charts.
'
' Retrieve the basic information.
thisProject = av.GetProject
theView = thisProject.FindDoc ("Maryland")
if (nil = theView) then
MsgBox.Error ("Unable to find the view document for
Maryland",
"")
exit
end

theTheme = theView.GetThemes.Get(0)
theVTab = theTheme.GetFTab
'
' Retrieve the required fields for charts.
fieldList1 = { theVTab.FindField("PUBLIC"),
theVTab.FindField("PRIVATE") }
fieldList2 = { theVTab.FindField("COST") }
nameField = theVTab.FindField("CNTY_NAME")
'
' Create the charts and set their properties.
pieChart = Chart.Make (theVTab, fieldList1)
pieChartWin = pieChart.GetWin
pieDisplay = pieChart.GetChartDisplay
barChart = Chart.Make (theVTab, fieldList2)
barChartWin = barChart.GetWin
barDisplay = barChart.GetChartDisplay
```

Scripts for Creating Charts 161

```
'
pieDisplay.SetType (#CHARTDISPLAY_PIE)
pieDisplay.SetStyle (#CHARTDISPLAY_VIEW_CUMULATIVE)
pieChart.SetName ("Student Enrollment")
pieChart.SetSeriesFromRecords (True)
pieChart.SetRecordLabelField (nameField)
pieChart.GetTitle.SetVisible (False)
if (pieDisplay.IsOK.Not) then
proceed = MsgBox.YesNo ("Pie chart may have an
inconsistency"
+NL+"Status:"++pieDisplay.GetStatus
+NL+"Do you want to continue?", "", False)
if (Not proceed) then
exit
end
end

'
barDisplay.SetType (#CHARTDISPLAY_BAR)
barDisplay.SetStyle (#CHARTDISPLAY_VIEW_SIDEBYSIDE)
barChart.SetName ("Expenditure")
barChart.SetSeriesFromRecords (True)
barChart.GetChartLegend.SetVisible (False)
barChart.GetYAxis.SetAxisVisible (False)
barChart.GetYAxis.SetLabelVisible (False)
barChart.GetXAxis.SetName ("Dollars")
barChart.GetXAxis.SetLabelVisible (True)
barChart.GetXAxis.SetMajorGridSpacing (1000)
barChart.GetTitle.SetName ("Per Student Enrolled In
Public Schools")
if (barDisplay.IsOK.Not) then
proceed = MsgBox.YesNo ("Bar chart may have an
inconsistency"
+NL+"Status:"++barDisplay.GetStatus
+NL+"Do you want to continue?", "", False)
```

```
if (Not proceed) then
exit
end
end
'
' Add the charts to the project and open them.
thisProject.AddDoc (pieChart)
thisProject.AddDoc (barChart)
' Close all documents first.
thisProject.CloseAll
theView.GetWin.Open
pieChartWin.Open
barChartWin.Open
av.TileWindows
```

Results of executing the above script are illustrated in the following figure.

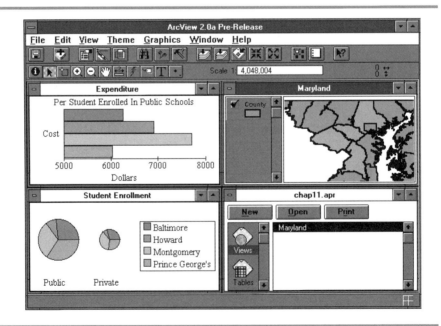

Education charts. (Coverage provided by Geographic Data Technology, Inc.)

Chart Document

A chart document displays the data in a table document as business graphics. Charts use the selected records and fields of a **VTab** to create graphics. VTabs are discussed in Chapter 9.

The **ChartDisplay**, **ChartLegend**, **Title**, **YAxis**, and **XAxis** classes make up the chart class, which can encompass any of the five or all of them. The five classes are manipulated to create the desired chart. Because charts are documents, you can open, close, and resize the window that holds a chart.

Creating a Chart

Because charts display data in tables, you must already have a table with attribute fields to create a chart. A chart is created by sending the **Make** request to the chart class. In the following example, the required parameters for the Make request are obtained, and the chart is created.

```
thisProject = av.GetProject
theView = thisProject.FindDoc ("Maryland")
theTheme = theView.GetThemes.Get(0)
theVTab = theTheme.GetFTab
fieldList1 = { theVTab.FindField("PUBLIC"),
theVTab.FindField("PRIVATE") }
'
' Create the chart.
pieChart = Chart.Make (theVTab, fieldList1)
```

The above approach is recommended for static applications in which the table or fields do not change. On the other hand, you can associate the **VTab** and open a chart property dialog box for the user to select the fields. In the following example, a chart is created after the user clicks on the OK button of the chart property dialog box.

```
thisProject = av.GetProject
theView = thisProject.FindDoc ("Maryland")
theTheme = theView.GetThemes.Get(0)
```

164 Programming Chart Documents

```
theVTab = theTheme.GetFTab
'
' Create the chart.
pieChart = Chart.MakeUsingDialog (theVTab)
```

The above approaches can be combined to display a chart property dialog box showing selected fields. Create the chart and then open the property dialog box as shown in the code segment below. The figure following the code segment shows the property dialog box.

```
' Create a chart.
pieChart = Chart.Make (theVTab, fieldList1)
' Open property dialog box.
pieChart.Edit
```

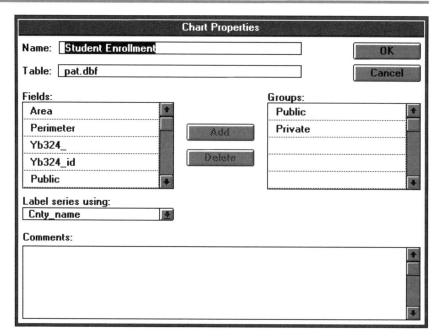

The chart property dialog box.

Understanding the ***ChartDisplay*** and ***ChartPart*** classes is important for programming chart documents. Because ChartDisplay holds the visual representation of the chart, manipulating a ChartDisplay object affects the representation. The following code line shows how to access the display object.

```
aChartDisplay = aChart.GetChartDisplay
```

ChartPart is an item in the chart such as the title or legend. ChartPart subclasses, including **Axis**, **Title**, and **ChartLegend**, are specific items on the chart and can be directly manipulated. The code segment below is an example of how to manipulate part of a chart.

```
barChart.GetTitle.SetName
("Per Student Enrolled In Public Schools")
```

Setting Chart Type and Style

Chart formats are defined by type and style. A suitable format for the underlying data can enhance presentation effectiveness. ArcView provides six chart types: area, column, bar, line, pie, and XY scatter charts. All types are offered in different styles. The following requests to a **ChartDisplay** object set the type and style of a chart.

```
aChartDisplay.SetType ( aChartType )
aChartDisplay.SetStyle ( aChartStyle )
```

SetType and **SetStyle** parameter enumerations are listed below.

Type enumerations (aChartType)

- #CHARTDISPLAY_AREA
- #CHARTDISPLAY_COLUMN
- #CHARTDISPLAY_BAR
- #CHARTDISPLAY_LINE
- #CHARTDISPLAY_PIE
- #CHARTDISPLAY_XYSCATTER

Style enumerations (aChartStyle)

- #CHARTDISPLAY_VIEW_CUMULATIVE
- #CHARTDISPLAY_VIEW_RELATIVE

❑ #CHARTDISPLAY_VIEW_SIDEBYSIDE

The following code segment is an example of setting chart type and style.

```
barChart = Chart.Make (theVTab, fieldList2)
barDisplay = barChart.GetChartDisplay
barDisplay.SetType (#CHARTDISPLAY_BAR)
barDisplay.SetStyle (#CHARTDISPLAY_VIEW_SIDEBYSIDE)
```

Through **ChartDisplay**, other visual properties can be set such as markers or symbols in line and scatter charts. Exercise care when setting ChartDisplay properties because not all combinations are valid. When properties are inconsistent, charts are not displayed. One way to prevent inconsistencies is to create the desired chart through ArcView. Then use **Get** requests to determine the properties set by ArcView. In the example below, the type and style of an existing chart are examined. The two figures following the script show the expected results.

```
myChart = av.GetProject.FindDoc ("Test Chart")
myChartDisplay = myChart.GetChartDisplay
MsgBox.Info ("Type:"
++myChartDisplay.GetType.AsString,
"Test Chart")
MsgBox.Info ("Style:"
++myChartDisplay.GetStyle.AsString,
"Test Chart")
```

The chart type display.

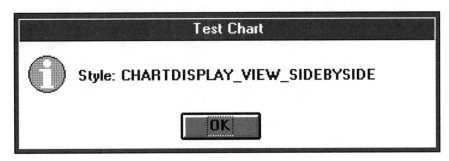

The chart style display.

Avenue provides a request to check for inconsistencies in a chart. The code segment below shows how to use the **IsOK** request to determine inconsistencies. If inconsistencies are found, the **GetStatus** request is used to identify them.

```
if (pieDisplay.IsOK.Not) then
proceed = MsgBox.YesNo ("Pie chart may have an
inconsistency"
+NL+"Status:"++pieDisplay.GetStatus
+NL+"Do you want to continue?", "", False)
if (Not proceed) then
exit
end
end
```

Setting Chart Properties

Chart properties are set by manipulating legend, title and axis objects. In addition, other properties such as name and data labels are accessed through the chart object. The following requests deal with the properties of a chart object.

```
groupByField = aChart.IsSeriesFromRecords
aChart.SetSeriesFromRecords (True)
```

168   Programming Chart Documents

A chart may be grouped by fields or records. In turn, series are set to records or fields. A group is a set of chart graphical elements that are brought together. A series is a set of chart graphical elements that are drawn once in each group. For example, in the "Student Enrollment" chart, "public" and "private" are groups while each county forms a series. The following code lines access and set the label of a group identified by *aGroupNumber*.

```
aGroupLabel = aChart.GetGroupLabel (aGroupNumber)
aChart.SetGroupLabel (aGroupNumber, aGroupLabel)
```

The *aGroupNumber* object refers to one group by its sequence number; zero refers to the first group. *aGroupLabel* is a string object.

```
aSeriesLabel = aChart.GetSeriesLabel (aSeriesNumber)
aChart.SetSeriesLabel (aSeriesNumber, aSeriesLabel)
```

The *aSeriesNumber* object refers to a series by its sequence number; zero refers to the first series. *aSeriesLabel* is a string object. If groups or series are from fields, the request to set a label also sets an alias for the field.

To access chart parts such as title or legend, the appropriate object must be retrieved. The following requests retrieve chart parts:

```
aChartLegend = aChart.GetChartLegend
aChartTitle = aChart.GetTitle
aChartXAxis = aChart.GetXAxis
aChartYAxis = aChart.GetYAxis
```

A chart part can be manipulated once you have its object. *XAxis* and *YAxis* are subclasses of *Axis*; Axis provides most of the requests for manipulating either axis. The code segment below is an example of setting properties for chart parts.

```
' Do not show the legend and Y axis.
barChart.GetChartLegend.SetVisible (False)
barChart.GetYAxis.SetAxisVisible (False)
barChart.GetYAxis.SetLabelVisible (False)
' Set X axis label to Dollars.
barChart.GetXAxis.SetName ("Dollars")
```

Chart Document   169

```
barChart.GetXAxis.SetLabelVisible (True)
' Set the chart title.
barChart.GetTitle.SetName
("Per Student Enrolled In Public Schools")
```

Various requests for each chart part are listed below:

```
' Set the visibility of an axis.
anAxis.SetAxisVisible (True)
' Set maximum and minimum values.
anAxis.SetBoundsMax (maxNumber)
anAxis.SetBoundsMin (minNumber)
anAxis.SetBoundsUsed (True)
' Set and display grid lines.
anAxis.SetMajorGridSpacing (aNumber)
anAxis.SetMajorGridVisible (True)
'
' Get or set the location of a chart legend.
aChartLoc = aChartLegend.GetLocation
aChartLegend.SetLocation (aChartLoc)
' Do not display a chart legend.
aChartLegend.SetVisible (False)
'
' Set the title.
aChartTitle.SetName (aString)
' Get or set the location of a chart title.
aChartLoc = aChartTitle.GetLocation
aChartTitle.SetLocation (aChartLoc)
' Display a chart title.
aChartLegend.SetVisible (True)
```

The *aChartLoc* enumerations follow:

```
#CHARTDISPLAY_LOC_BOTTOM
#CHARTDISPLAY_LOC_LEFT
#CHARTDISPLAY_LOC_RELATIVE
```

170   Programming Chart Documents

```
#CHARTDISPLAY_LOC_RIGHT
#CHARTDISPLAY_LOC_TOP
```

Chart is a subclass of *Doc*. A chart object inherits all the valid requests for Doc. Some of the most commonly used requests appear in the code segment below.

```
' Document window is needed to open or
' close a document.
aChartWin = aChart.GetWin
aChartWin.Open
aChartWin.Close
'
' Document name appears in the project window and
' document title bar.
aChart.SetName ("My Chart")
```

Working with Data Elements

In most applications, tabular data is used to create a chart. However, in some cases you may wish to access data from a chart. Once a chart is created in an application, the user can click on a customized pushbutton to calculate a series of complex statistics. For example, in a redistricting application a bar chart might show the population's racial composition for an arbitrary set of districts. The user then clicks on a pushbutton to compute the standard deviation for all districts in the chart.

The following data elements can be accessed: records in the chart, fields used in the chart, a record identified by the user, and a record with a matching string of characters. The data elements are described in the following sections.

Accessing Records

The chart document displays selected records in a *VTab*. If no records are selected, all records are displayed in a chart. To access the records

Working with Data Elements   171

from a chart, you need to get to its VTab and then to selected rows in the VTab. The code segment below demonstrates this approach.

```
theVTab = myChart.GetVTab
selectedRecords = theVTab.GetSelection
if (selectedRecords.Count = 0) then
' No records were selected.
' Therefore, all records are shown in the chart.
else
for each recNumber in selectedRecords
' The record referenced by the recNumber variable
' is part of the chart.
end
end
```

The **GetSelection** request returns a bit map object. A bit map is an ordered, fixed size list of Boolean objects. Each Boolean object represents a record with matching index number in the VTab. If the Boolean object is set to True, the record is selected. A list of useful requests for a bit map object appears below.

```
' Clear all bits by assigning a False value.
aBitmap.ClearAll
'
' Toggle True and False bits.
aBitmap.Not
'
' Set all bits to True.
aBitmap.SetAll
'
' Return number of bits that are set.
countSet = aBitmap.Count
'
' Return the size of a bit map.
size = aBitMap.GetSize
'
```

172   Programming Chart Documents

```
' Evaluate a bit map at a bit location.
isSet = aBitmap.Get (aBitLocation)
```

Accessing Fields

A list of fields used in a chart is returned by the ***GetFields*** request to a chart object. The code segment below shows how to check if the *POP18* field is used in a chart.

```
fieldsList = myChart.GetFields
for each aField in fieldsList
If (aField.GetName = "POP18") then
' The field has been found.
end
end
```

Identifying a Record

ArcView provides an ***Identify*** tool button for chart documents. With the use of this tool, ArcView displays the fields of a particular record which pertain to a chart data element. The Identify tool button can also be used to perform other tasks with the identified record. In the code segment below, the ***GetRecordFromClick*** request accepts a mouse click from the user. If the user clicks on a data element, the request returns the associated record number.

```
theVTab = myChart.GetVtab
recNumber = myChart.GetUserRecord
if (recNumber >= 0) then
population = theVTab.ReturnValue
(popField,recNumber)
end
```

Finding a Record by Matching a String

The ***Find*** request can search for a matching string in any of the character fields of the associated ***VTab***. If a record is found, the request returns a record number. A nil object is returned if a record is not found. The format for this request is shown below.

```
recNumber = myChart.Find (aString)
```

Printing Charts

Printing a chart is simple. The only task involved is sending the ***Print*** request to a chart object as shown in the code line below.

```
myChart.Print
```

Chapter 11

Programming Layout Documents

Users prepare hardcopy maps in ArcView by using layout documents. A layout document is a map which can contain views, tables, charts, imported graphics, and graphic primitives. It can also contain cartographic elements such as a scale bar and north arrow.

Avenue can automate preparation of a layout document. You can write an Avenue script to guide a user through generating a custom map or producing a standard layout. For example, map standards in your organization may require that the north arrow appear above the scale bar at the bottom center of the printed page. Avenue statements can provide the flexibility to pick from different styles of north arrow or scale bar, while placing them at bottom center of the sheet. Another situation might be that your organization uses only a specific style of north arrow regardless of its location on the sheet. An Avenue script can provide flexibility of location while forcing a predefined style of north arrow.

This chapter focuses on how you can use Avenue to create and modify a layout document. A script will be created and then associated to a

176 Programming Layout Documents

customized pushbutton of a view document interface. Clicking the pushbutton generates a layout as shown in the following figure.

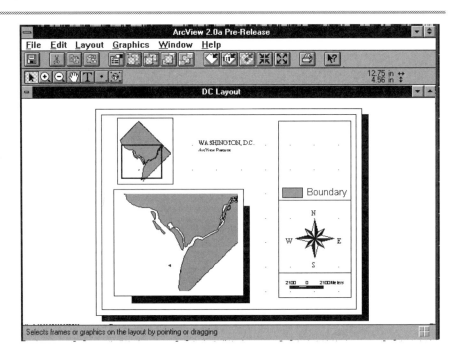

Layout document generated by an Avenue script. (Coverage provided by Geographic Data Technology, Inc.)

Layout Document Script

The entire script follows. Each segment of the script is explained before the code listing.

```
' Create_a_Map
' This script is attached to a customized
' pushbutton in the view document interface.
' Executing this script will create a standard
' layout document from the current view.
```

Layout Document Script 177

```
'
' First create a new layout.
'
thisProject = av.GetProject
thisView = av.GetActiveDoc
stdLayout = Layout.Make
stdLayout.SetName (thisView.GetName++"Layout")
stdLayoutDpy = stdLayout.GetDisplay
stdLayoutGL = stdLayout.GetGraphics
'
' Set the layout page properties to
' landscape 11 x 8.5 with all around
' 0.00 inch margin.
'
marginRect = Rect.MakeXY ( 0.0, 0.0, 0.0, 0.0)
stdLayoutDpy.SetUnits (#UNITS_LINEAR_INCHES)
stdLayoutDpy.SetMargin (marginRect)
stdLayoutDpy.SetMarginVisible (True)
stdLayoutDpy.SetPageSize (Point.Make(11.0,8.5))
stdLayoutDpy.SetGridActive (False)
stdLayoutDpy.SetGridVisible (True)
stdLayoutDpy.SetGridMesh (Point.Make(1.0,1.0))
'
' Requests that place graphic objects on the
' layout require X and Y coordinates. The
' coordinates are based on the Display frame
' and not the PageDisplay.
' In order to use the coordinates of
' PageDisplay, the
' lower left X and Y for the PageDisplay is
' obtained. In the next code line this
' coordinate is stored in
' the oPt (origin point) object.
'
oPt = stdLayoutDpy.ReturnMarginRect.ReturnOrigin
```

178   Programming Layout Documents

```
'
' Add a title and a subtitle to the page.
' Use Times Roman font and sizes of 24 and 18.
'
titleSymbol = TextSymbol.Make
titleSymbol.SetFont (
Font.Make("Times New Roman","Normal") )
titleSymbol.SetSize (24) 'size is in points
subtitleSymbol = titleSymbol.Clone
subtitleSymbol.SetSize (18)
allTitles = MsgBox.MultiInput (
"Please enter","",
{"Title:","Subtitle:"},
{"",""})
if (allTitles.Count 0) then
pTitle = GraphicText.Make (
allTitles.Get(0), oPt+Point.Make(4.25,7.0) )
pTitle.SetSymbols ({titleSymbol})
pTitle.SetAngle (0)
stdLayoutGL.Add (pTitle)
subtitle = GraphicText.Make (
allTitles.Get(1), oPt+Point.Make(4.25,6.75) )
subtitle.SetSymbols ({subtitleSymbol})
subtitle.SetAngle (0)
stdLayoutGL.Add (subtitle)
end

'
' Add view frames. The primary view frame
' shows the active view at its current extent,
' while the locator view frame shows the
' active frame zoomed to its extent with
' a rectangle depicting the area of primary view.
'
' The primary view is placed at point 1,1 with
```

Layout Document Script 179

```
' a shadow box.
'
vFill = RasterFill.Make
vFill.SetColor (Color.GetWhite)
vFill.SetStyle (#RASTERFILL_STYLE_SOLID)
vRect = Rect.Make (
oPt+Point.Make (1.0,1.0),
Point.Make (6.0,4.0) )
vFrame = ViewFrame.make (vRect)
vFrame.SetSymbol (vFill)
vFrame.SetView (thisView,True)
vFrame.SetScalePreserved (False)
stdLayoutGL.Add (vFrame)
'
' Add a shadow box to the primary view.
'
sFill = vFill.Clone
sFill.SetColor (Color.GetBlack)
sRect = vFrame.GetBounds
shadowGr = GraphicShape.Make (sRect)
shadowGr.Offset (Point.Make(0.5,-0.5))
shadowGr.SetSymbol (sFill)
stdLayoutGL.UnselectAll
shadowGr.SetSelected (True)
stdLayoutGL.Add (shadowGr)
stdLayoutGL.MoveSelectedToBack
stdLayoutGL.UnselectAll
'
' Add the locator map at 1,5.5 inches.
' The locator map is based on a cloned view that
' displays the view's full extent.
'
lRect = Rect.Make (
oPt+Point.Make(1.0,5.5),
Point.Make (3.0,2.5) )
```

180   Programming Layout Documents

```
lFrame = ViewFrame.Make (lRect)
lFrame.SetSymbol (vFill)
fullView = thisView.Clone
fullView.GetDisplay.SetExtent
(fullView.ReturnExtent)
fullView.SetName("Full View")
thisProject.AddDoc (fullView)
lFrame.SetView (thisView, True)
lFrame.SetScalePreserved (False)
stdLayoutGL.Add (lFrame)
'
' Draw a box in the locator map to show
' the extent of primary view. This box is
' actually drawn on the view display and
' seen on the layout document.
'
boxFill = vFill.Clone
boxFill.SetStyle (#RASTERFILL_STYLE_EMPTY)
boxFill.SetOutlined (True)
boxFill.SetOlColor (Color.GetBlack)
boxFill.SetOlWidth (4)
vFrameExtent = vFrame.GetMapDisplay.ReturnExtent
lBox = GraphicShape.Make (vFrameExtent)
lBox.SetSymbol (boxFill)
fullViewGL = fullView.GetGraphics
fullViewGL.Add (lBox)
'
' Draw boxes to hold the scale bar,
' north arrow, and legend. AddBatch is
' used to add the remaining items to
' the layout.
'
aPen = BasicPen.Make
aPen.SetColor (Color.GetBlack)
infoBox = Polygon.Make ( { {
```

Layout Document Script 181

```
oPt+Point.Make (7.50,0.75),
oPt+Point.Make (7.50,8.00),
oPt+Point.Make (10.5,8.00),
oPt+Point.Make (10.5,0.75) } } )
infoBoxGr = GraphicShape.make (infoBox)
infoBoxGr.SetSymbol (aPen)
stdLayoutGL.AddBatch (infoBoxGr)
'
line1 = Line.Make (
oPt+Point.Make (7.50,1.75),
oPt+Point.Make (10.5,1.75) )
line1Gr = GraphicShape.Make (line1)
line1Gr.SetSymbol (aPen)
stdLayoutGL.AddBatch (line1Gr)
'
line2 = Line.Make (
oPt+Point.Make (7.50,4.75),
oPt+Point.Make (10.5,4.75) )
line2Gr = GraphicShape.Make (line2)
line2Gr.SetSymbol (aPen)
stdLayoutGL.AddBatch (line2Gr)
'
' Add scale bar.
'
sbRect = Rect.Make (
oPt+Point.Make (7.8,1.0),
Point.Make (2.50,0.50) )
sbFrame = ScalebarFrame.Make (sbRect)
sbFrame.SetUnits (#UNITS_LINEAR_KILOMETERS)
sbFrame.SetStyle (#SCALEBARFRAME_STYLE_ALTFILLED)
sbFrame.SetViewFrame (vFrame)
stdLayoutGL.AddBatch (sbFrame)
'
' Add north arrow.
'
```

182 Programming Layout Documents

```
naRect = Rect.Make (
oPt+Point.Make (7.9, 1.8),
Point.Make (2.75,3.0) )
naGr = NorthArrow.Make (naRect)
' Retrieve a predefined north arrow from
' the north.def file.
northArrowFile = File.Make
("\win32app\arcview\etc\north.def")
northArrowODB = ODB.Open (northArrowFile)
if (nil = northArrowODB) then
MsgBox.Error
("Unable to open north.def object database",
"")
exit
end

northArrowList = northArrowODB.Get(0)
anArrow = northArrowList.Get(0)
naGr.SetArrow (anArrow)
stdLayoutGL.AddBatch (naGr)
'
' Add the legend.
'
lgRect = Rect.Make (
oPt+Point.Make (7.75,4.85),
Point.Make (5.65,0.30) )
lgFrame = LegendFrame.Make (lgRect)
lgFrame.SetViewFrame (vFrame)
stdLayoutGL.AddBAtch (lgFrame)
'
' End the AddBatch.
'
stdLayoutGL.EndBatch
'
' Open the layout document.
```

Layout Documents   183

```
stdLayoutWin = stdLayout.GetWin
stdLayoutWin.Open
stdLayoutDpy.ZoomToPage
```

The preceding script should be associated with a customized push-button of the view document interface. The following figure shows the **Customize** dialog box for this button.

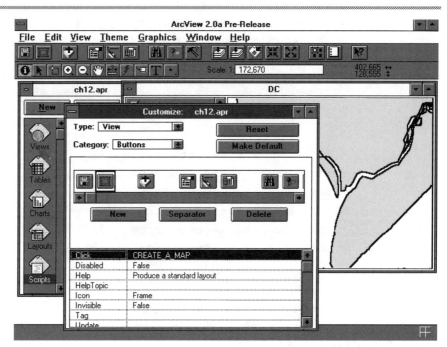

Customized pushbutton for CREATE_A_MAP script.

Layout Documents

Layout is a specialized ***Doc*** class, and can be manipulated in the same way as other documents. This section focuses on how to create a new layout, set properties and display the layout.

184   Programming Layout Documents

Creating a Layout

As seen in the following Avenue statement, creating a layout document requires only a ***Make*** request sent to the ***Layout*** class. The newly created layout document is automatically added to the project file.

```
stdLayout = Layout.Make
```

Similar to other documents, a layout has display and window objects. Another important object is the graphic list. The graphic list for a layout document maintains all elements added to the document. For example, the list holds title text, the north arrow, and views. The graphic list is discussed in greater detail later in this section.

Retrieval of the display, window and graphic objects is recommended after creating the layout document. The code segment below shows how the objects are retrieved.

```
' Retrieve the layout display object.
stdLayoutDpy = stdLayout.GetDisplay
' Retrieve the layout document window.
stdlayoutWin = stdLayout.GetWin
' Retrieve the layout graphic list.
stdLayoutGL = stdLayout.GetGraphics
```

The display object is required to manipulate document contents, such as in pan, zoom and draw. The window object is for document management, e.g., open and maximize.

Layout Display

Once the layout is created, you can access its window object in order to display the layout and ask user approval before sending it to the printer. In the following code segment, the layout is opened, and the user is asked if the layout should be printed.

```
' Retrieve the window object.
myLayoutWin = myLayout.GetWin
' Open layout if closed.
if (myLayoutWin.IsOpen.Not) then
```

```
myLayoutWin.Open
end

' Maximize layout.
myLayoutWin.Maximize
' Ask user whether to print.
goAhead = MsgBox.YesNo ("Print this layout?",
"My Application", True)
if (goAhead) then
myLayout.Print
else
MsgBox.Info ("Print canceled.","My Application")
end

' Close the layout document.
myLayoutWin.Close
```

Setting Layout Properties

The main properties of a layout document are related to printer page layout components such as page size and margins. Assuming that the printer has been configured in ArcView, you can simply set the page size and margins to the printer's default values. Use the following requests to obtain the printer default values set in the current configuration.

```
' Both requests are made to the
' layout display object.
myLayoutDpy.SetUsingPrinterPageSize (True)
myLayoutDpy.SetUsingPrinterMargins (True)
```

You can directly set page properties by sending the following requests to the layout display object.

```
' Set page units to inches.
stdLayoutDpy.SetUnits (#UNITS_LINEAR_INCHES)
' Setting to landscape letter size.
' Set page by sending a point equivalent
' to the page size.
```

186   Programming Layout Documents

```
stdlayoutDpy.SetPageSize (Point.Make(11,8.5))
' Setting margins to 0.25 inches all around.
' Set the margin by sending a rectangle
' equivalent to the margins.
marginRect = Rect.MakeXY (0.25, 0.25, 0.25, 0.25)
stdlayoutDpy.SetMargin (margin.Rect)
' Verify that margin lines are displayed.
stdLayoutDpy.SetMarginVisible (True)
```

In the preceding code segment, the page measurement unit is set by the enumeration value, *#UNITS_LINEAR_INCHES*. Additional unit enumerations appear below.

- ❏ #UNITS_LINEAR_FEET
- ❏ #UNITS_LINEAR_YARDS
- ❏ #UNITS_LINEAR_MILES
- ❏ #UNITS_LINEAR_MILLIMETERS
- ❏ #UNITS_LINEAR_CENTIMETERS
- ❏ #UNITS_LINEAR_METERS
- ❏ #UNITS_LINEAR_KILOMETERS
- ❏ #UNITS_LINEAR_NAUTICALMILES
- ❏ #UNITS_LINEAR_DEGREES
- ❏ #UNITS_LINEAR_PRJMETERS

Grid is another page property. A grid is often useful when a user is interactively creating the layout document because it provides a visual aid for placing layout components. The grid snap also ensures that components are placed in exact locations. However, when preparing a layout through Avenue, a grid system is not required because components can be located at precise locations. The grid can nonetheless be displayed to provide a visual aid as to how each component is placed in relation to another. In the following code segment, the layout displays a one-inch grid mesh without snapping.

```
' Grid properties are set by sending
' requests to the layout display object.
'
```

```
' Setting the grid mesh to one-inch intervals:
' set the mesh in x and y directions by sending
' an equivalent point.
stdLayoutDpy.SetGridMesh (Point.Make(1,1))
' Display the grid.
stdLayoutDpy.SetGridVisible (True)
' Turn off the snap.
stdLayoutDpy.SetGridActive (False)
```

Because layout is a type of document, it has properties similar to other documents. The document name, one of the most useful properties, appears on the title bar of a document window. Document name can also be used to search for a document. In the following code segment, a layout document is named by adding the word "Layout" to the primary view name.

```
stdLayout.SetName (thisView.GetName++"Layout")
```

Using the Graphic List

Placing a component on the layout results in adding it to the layout graphic list. The list is used in accessing and manipulation of all layout components.

The **GraphicList** class is a specialization of the **List** class. The GraphicList class holds the graphics associated with document display. Whenever a document display is redrawn, the graphic list is used to place all the graphics. The graphics in the list are comprised of mathematical representations of shapes, and become objects of the **GraphicShape** class. GraphicShape provides the algorithm that allows each shape to draw itself.

A simple sequence of adding graphics to a document involves the following steps: retrieve the mathematical representation of the graphic shape, add it to the graphic list, and trigger a redraw of the document. In the following code segment, the sequence is illustrated, and a polygon is added to the layout document.

```
' Get the layout graphic list by
' sending the GetGraphics request
```

188 Programming Layout Documents

```
' to the layout document.
stdLayoutGL = stdLayout.GetGraphics
' Create a shape by sending a
' Make request to a shape class.
infoBox = Polygon.Make ( { {
oPt+Point.Make (7.50,0.75),
oPt+Point.Make (7.50,8.00),
oPt+Point.Make (10.5,8.00),
oPt+Point.Make (10.5,0.75) } } )
' Generate the mathematical
' representation of the shape
' by sending a Make request to
' the GraphicShape class.
infoBoxGr = GraphicShape.make (infoBox)
' Add the graphic shape to the
' graphic list and cause a redraw.
stdLayoutGL.Add (infoBoxGr)
stdLayout.Invalidate
```

When several graphic shapes are added to a graphic list, it is frequently more efficient to use the ***AddBatch*** request in conjunction with the ***EndBatch*** request. These requests are shown in the statements below.

```
' Simultaneously add several line graphic
' shapes to the graphic list.
myLayoutGL.AddBatch (line1Gr)
myLayoutGL.AddBatch (line2Gr)
myLayoutGL.AddBatch (line3Gr)
myLayoutGL.EndBatch
```

The following code segment erases all layout components.

```
myLayoutGL = myLayout.GetGraphics
' Select all graphics.
myLayoutGL.SelectAll
```

Layout Documents 189

```
' Erase all selected graphics.
myLayoutGL.ClearSelected
```

An object from the ***GraphicText*** class is used for adding text to the layout document. In the following code segment, this process is illustrated while a title is added to the layout document.

```
' Get the layout graphic list
' by sending GetGraphics request
' to the layout document.
stdLayoutGL = stdLayout.GetGraphics
' Get the title from the user.
allTitles = MsgBox.MultiInput (
"Please enter","",
{"Title:","Subtitle:"},
{"",""})
if (allTitles.Count 0) then
' Generate the mathematical
' representation of the text
' by sending a Make request to
' the GraphicText class.
pTitle = GraphicText.Make (
allTitles.Get(0), locationPoint)
' Add the graphic text to the list.
stdLayoutGL.Add (pTitle)
end
```

The ***GraphicShape*** and ***GraphicText*** classes create the graphics for shapes and text to add to the layout graphic list. Graphical frames are required to add views, charts, or tables. Frames, which can be described as containers for views or other layout components, are discussed in the following sections.

190   Programming Layout Documents

Framing a View Document

Views are placed on a layout document through objects of the **View-Frame** class. A view frame is created by sending the **Make** request to the ViewFrame class. As a parameter, this request requires a rectangle which represents the frame boundary. An existing view document can also be associated to the frame. If a view document is not associated, then the frame is displayed as an empty view frame. In the following code segment, a view frame is created, its properties are set, and the frame is added to the graphic list.

```
' Add a view frame to stdLayout document.
'
' Set the frame boundary in vRect. Layout
' may change the frame boundary based on
' the width and height of the associated view.
vRect = Rect.Make (
oPt+Point.Make (1.0,1.0),
Point.Make (6.0,4.0) )
' Create the view frame.
vFrame = ViewFrame.make (vRect)
' Assign a view to the frame. The second
' parameter sets the live link; if set to False,
' view frame becomes a snap shot of the view
' document.
vFrame.SetView (thisView,True)
' By preserving the scale, frame displays the
' view at its current scale.
vFrame.SetScalePreserved (False)
' Add the view frame to the graphic list.
stdLayoutGL.Add (vFrame)
' Redraw the layout.
stdLayout.Invalidate
```

The layout document may change the size of the view frame based on the width and height of the associated view. The **GetBounds** request

Framing a View Document   191

can be used to obtain the new size. This request is useful when placing another object on the layout based on the view frame. For example, if you want to place a shadow box under the view frame, you need the exact frame size. In the following code segment, a shadow box for the view frame is added to the layout.

```
' Add a shadow box to the vFrame object.
'
' First, create a solid black fill symbol.
sFill = RasterFill.Make
sFill.SetStyle (#RASTERFILL_STYLE_SOLID)
sFill.SetColor (Color.GetBlack)
' Set the shadow box size to the same size as
' vFrame. GetBounds returns a rectangle.
sRect = vFrame.GetBounds
' Create a mathematical representation of a
' rectangle.
shadowGr = GraphicShape.Make (sRect)
' Associate the graphic shape with the layout
' display.
shadowGr.SetDisplay (stdLayoutDpy)
' Move the shadow box half inches to the right
' and down.
shadowGr.Offset (Point.Make(0.5,-0.5))
' Set the shadow box to a black solid fill.
shadowGr.SetSymbol (sFill)
' Verify that only the shadow box
' in the graphic list is selected.
stdLayoutGL.UnselectAll
shadowGr.SetSelected (True)
' Add shadow box to the layout, and then
' move it to the back of the view frame.
stdLayoutGL.Add (shadowGr)
stdLayoutGL.MoveSelectedToBack
stdLayoutGL.UnselectAll
```

192   Programming Layout Documents

Once a **ViewFrame** object is placed on the layout, the associated scale bar and legend can be added.

Adding a Scale Bar

A scale bar can be drawn by associating a **ScaleBarFrame** object to a view frame and adding it to the graphic list. Send the **Make** request to the class to create a scale bar frame object. The Make request requires a rectangle that sets the scale bar's boundary. The following code segment is an example of how to add a scale bar to a view.

```
' Add a scale bar.
'
' Create a rectangle defining scale
' bar boundary.
sbRect = Rect.Make (
oPt+Point.Make (8.25,1.25),
Point.Make (2.50,0.50) )
' Create the scale bar object.
sbFrame = ScalebarFrame.Make (sbRect)
' Set scale bar properties.
sbFrame.SetUnits (#UNITS_LINEAR_KILOMETERS)
sbFrame.SetStyle (#SCALEBARFRAME_STYLE_ALTFILLED)
' Associate the bar with the vFrame.
sbFrame.SetViewFrame (vFrame)
' Add the bar to the graphic list.
stdLayoutGL.Add (sbFrame)
```

Several scale bar styles are available through the bar style enumeration. The styles are listed below.

- ❏ #SCALEBARFRAME_STYLE_ALTFILLED
- ❏ #SCALEBARFRAME_STYLE_FILLED
- ❏ #SCALEBARFRAME_STYLE_HOLLOW
- ❏ #SCALEBARFRAME_STYLE_LINED
- ❏ #SCALEBARFRAME_STYLE_TEXT

Additional properties of a scale bar can be set through Avenue. These properties are shown in the following code lines.

```
' Set the number of divisions in the left
' interval of the scale bar.
myScaleBar.SetDivisions (aNumber)
'
' Set the interval of the scale bar.
myScaleBar.SetInterval (aNumber)
```

Adding Legends

The view document legend can be displayed on the layout inside a legend frame. An object of the **LegendFrame** class is created by sending a **Make** request. The Make request requires a rectangle corresponding to the boundary of the legend frame. A legend frame must be associated to a view frame. The following code segment shows how to add a legend to the layout.

```
' Add a legend.
'
' Create a rectangle for the frame boundary.
lgRect = Rect.Make (
oPt+Point.Make (7.75,4.85),
Point.Make (5.65,0.30) )
' Create a legend frame object.
lgFrame = LegendFrame.Make (lgRect)
' Associate the legend with the view frame.
lgFrame.SetViewFrame (vFrame)
' Add the legend frame to the graphic list.
stdLayoutGL.Add (lgFrame)
```

Adding a North Arrow

The **NorthArrow** class is a specialized class of **GraphicGroup**. NorthArrow objects are limited to graphic shapes and graphic text. You may create your own north arrow by creating shape and text graphics

194   Programming Layout Documents

and adding them to the NorthArrow graphic group. However, ArcView comes with an ODB (object database) file that defines several types of north arrows. ODBs are discussed in Chapter 12. The following code segment shows how to access the *north.def* file in the *etc* directory to retrieve a NorthArrow object.

```
' Several north arrow shapes are defined
' in the north.def object database file.
northArrowFile = File.Make
("\win32app\arcview\etc\north.def")
northArrowODB = ODB.Open (northArrowFile)
' The first object in the ODB is a list of
' stored objects.
northArrowList = northArrowODB.Get(0)
anArrow = northArrowList.Get(0)
naGr.SetArrow (anArrow)
stdLayoutGL.Add (naGr)
```

Framing a Table or Chart Document

A table or chart document is placed on a layout by adding a **DocFrame** object to the layout graphic list. The following code segment shows how to create a DocFrame object and add it to the graphic list.

```
' Document frames are created within
' rectangle boundaries.
tableRect = Rect.MakeXY (x1, y1, x2, y2)
tableFrame = DocFrame.Make (tableRect, Table)
' Associate frame with the table document.
tableFrame.SetFramedDoc (myTable)
' Add frame to the layout graphic list.
myLayoutGL.Add (tableFrame)
'
' A chart document frame is created in
' a similar way.
```

Framing a Picture — 195

```
chartRect = Rect.MakeXY (x1, y1, x2, y2)
chartFrame = DocFrame.Make (chartRect, Chart)
' Associate frame with the chart document.
chartFrame.SetFramedDoc (myChart)
' Add frame to the layout graphic list.
myLayoutGL.Add (chartFrame)
```

The layout does not display the contents of a table or chart document unless the table or chart document is open. A **DocFrame** object is a type of frame that holds a table or chart on the layout. The following properties can be set for a DocFrame object:

```
' Set presentation quality.
myDocFrame.SetQuality (#FRAME_QUALITY_DRAFT)
' The other enumeration for presentation
' quality is #FRAME_QUALITY_PRESENTATION.
'
' Set the frame refresh property.
myDocFrame.SetRefresh (#FRAME_REFRESH_WHENACTIVE)
' The other enumeration for the refresh
' property is #FRAME_REFRESH_ALWAYS.
```

Framing a Picture

Contents of a graphic file can be placed on the layout by adding a **PictureFrame** object to the graphic list. A graphic disk file such as your organization's logo or a property photo can be imported into a picture frame. In the following code segment, a rectangle area for the company logo is created.

```
logoRect = Rect.MakeXY (x1, y1, x2, y2)
' Create a picture frame within the logoRect
' boundaries.
logoFrame = PictureFrame.Make (logoRect)
' Import the graphic file into the frame.
```

196   Programming Layout Documents

```
logoFrame.SetFileName ("LOGO.TIF".AsFileName)
' Add file to the graphic list.
myLayoutGL.Add (logoFrame)
```

The presentation quality and refresh properties for a picture frame can also be set. The appropriate requests are shown in the following code segment.

```
logoFrame.SetQuality (qualityEnum)
' The qualityEnum could have one of the
' following values: #FRAME_QUALITY_DRAFT,
' or #FRAME_QUALITY_PRESENTATION.
'
logoFrame.SetRefresh (refreshEnum)
' The refreshEnum could have one of the
' following values:
' #FRAME_REFRESH_WHENACTIVE,
' or #FRAME_REFRESH_ALWAYS.
```

The user could also be allowed to select the graphic file by browsing through directories. In the following code segment, a picture frame is created but the user selects the graphic file. The picture frame is selected prior to placing it on the layout so that the user can immediately manipulate its size and location.

```
photoRect = Rect.MakeXY (x1, y1, x2, y2)
photoFrame = PictureFrame.Make (photoRect)
if (photoFrame.Edit (myLayoutGL)) then
' Verify that nothing is selected.
myLayoutGL.UnselectAll
' Select the picture frame.
photoFrame.SetSelected (True)
' Add the frame to the layout graphic list.
myLayoutGL.Add (photoFrame)
end
```

The *Edit* request is used to set a frame's properties, including the graphic file for import. The property dialog box as shown in the

Framing a Picture 197

following figure contains an OK and a **Cancel** button. The Edit request returns True if OK is clicked. If the user does not click on OK, the request returns False.

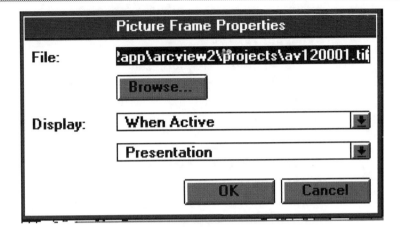

Picture frame property dialog box.

ArcView supports the following graphic formats:
- Postscript
- Encapsulated Postscript
- Windows Bitmap
- Windows Metafile
- X-Bitmap
- Mac Paint
- Macintosh Pict
- Sun Raster
- GIF
- TIFF
- Nexpert Object Image
- ERDAS
- Run Length Compression
- Band Interleave

198   Programming Layout Documents

Printing the Layout

Sending a layout to a printer is a simple task. The following code line prints a layout on the default printer.

```
myLayout.Print
```

The **Printer** class represents a hardware independent hardcopy output device. Avenue statements can be used for printer setup, or to obtain setup information. The following code lines show how to access and configure the default printer.

```
' The "The" request to the
' printer class retrieves the default
' printer object.
aPrinter = Printer.The
' Open the setup dialog box.
aPrinter.SetUp
' Check to see if printer is on-line.
if (aPrinter.IsReady.Not) then
MsgBox.Warning ("Printer off-line.", "")
end
' Print to a file.
aPrinter.SetFileName (aFileName)
' Stop printing to a file.
aPrinter.SetFileName ("")
```

Chapter

12

Application Installation

Users can access your application either by opening a predefined project file or by loading it through default settings. The project file of a frequently used application should be configured as the default project for ArcView. However, if many applications are frequently used, each application should be maintained in a separate project file, and a menu system established in the default project file to reach each application.

Regardless of the approach you take, you need to safeguard your application by protecting its scripts. This chapter focuses on protecting scripts, and how to make an application the default project file.

200   Application Installation

Protecting Your Scripts

While you may wish to allow users to see or copy your scripts, you should prevent users from modifying the scripts. Among other things, users may inadvertently crash the entire application by changing a script.

ArcView provides two levels of protection: embedding and encryption. Before undertaking either level of protection, verify that the script is compiled and free of errors.

Embedding means that the script is part of the project file but is not displayed on the project window. Meanwhile, users can copy the script from the **Script Manager** dialog box. A script can be embedded through ArcView menu options or Avenue statements. If you are working with many scripts, setting up Avenue codes for embedding is more efficient. In the following code segment, a script is displayed on the project window, embedded and removed from the project window, and then unembedded and returned to the project window.

```
' Get the script object.
mySEd = av.GetProject.FindDoc ("App Script")
myScript = mySEd.GetScript
if (nil = myScript) then
MsgBox.Error ("Script is not compiled.", "")
exit
end

' Embed the script by first adding it to the
' project as a new script, and then removing
' the old script from the project.
myEmbScript = myScript.Clone
av.GetProject.AddScript (myEmbScript)
av.GetProject.RemoveDoc (mySEd)
'
' Return the embedded script to the
' project window.
myScript = av.FindScript ("App Script")
' Create a script editor document.
```

Protecting Your Scripts 201

```
mySEd = SEd.Make
' Add the script editor document to
' the project window.
av.GetProject.AddDoc (mySEd)
' Load the script into the editor.
mySEd.SetSource (myScript.AsString)
```

Encryption is an irreversible process which writes a cipher and encrypts the contents of a script. The script is listed in the **Script Manager** dialog box, but its contents cannot be copied or viewed. Once a script is encrypted, it cannot be decrypted. Therefore, you should always make a copy of the source code for yourself prior to encryption.

Scripts can be encrypted separately or simultaneously. The following code segment shows how to encrypt a script. This process creates a new script object that does not replace the source code. As shown in the following code segment, the new encrypted script must be named. The script name can then be viewed in the **Script Manager** dialog box.

```
' Start with the script object.
myScript = av.FindScript ("App Script")
' Create a new encrypted script object and
' give it a name.
myEncScript = EncryptedScript.MakeFromScript
(myScript)
myEncScript.SetName
(myScript.GetName+".ENC")
' Embed it into the project file.
av.GetProject.AddScript (myEncScript)
' Save the project file.
av.GetProject.Save
```

To simultaneously encrypt all scripts, make the ***EncryptScripts*** request to the project object. EncryptScripts encrypts the source for all scripts and then removes them from the project window. This process is irreversible, and source codes are lost. Consequently, you should copy the entire project or the scripts prior to encryption. Use this request with

202   Application Installation

caution. In the following code segment, the project is saved and then encrypted as a new project.

```
thisProject = av.GetProject
' Save the project with source codes.
thisProject.Save
' Get a new name for the project with
' encrypted codes.
encFileName = FileDialog.Put
("","*.apr","Save Encrypted Project As")
if (nil  encFileName) then
thisProject.SetFileName (encFileName)
thisProject.EncryptScripts
thisProject.Save
MsgBox.Info ("Encrypted project saved as"++
encFileName.GetBaseName)
end
```

The **EncryptScripts** request encrypts all project scripts except those beginning with a period, such as *.myScript* or *.Script1*. Therefore, prior to making the request, place a period at the beginning of the script file names that you do not wish to encrypt.

Distributing Objects

Objects can be saved to disk files. This feature allows you to send updates for your application, or to distribute value added objects outside a project file. ArcView stores objects in object database files. The **Odb** class in Avenue, which represents an object database, provides a file-based storage and retrieval system for ArcView objects. In the following code segment, a script is stored in an object database.

```
' Ask user to provide a new file name.
aFileName = FileDialog.Put
("","*.odb","Save Object As")
' Create the file and Odb object.
```

```
myOdb = Odb.Make (aFileName)
' Add the script to the object database.
myScript = av.FindScript ("App Script")
myOdb.Add (myScript)
' Save the object database.
worked = myOdb.Commit
if (worked.NOT) then
MsgBox.Error ("Unable to add"++
myScript.GetName++"to"++
aFileName.GetBaseName)
end
```

More than one object can be added to an object database. Following all the **Add** requests, the **Commit** request is sent to the **Odb** object only once. The Commit request saves the objects to the database file.

To retrieve an object from the object database, you need to open the database file and then use the **Get** request. This process is shown in the following code segment, where the *script.odb* file is opened and all script objects extracted.

```
' Create an ODB object by opening an
' existing object database.
myOdb = Odb.Open ("script.odb".AsFileName)
' Create a list to hold the scripts.
scriptList = {}
' Loop through all scripts in ODB
' and add it to the list.
for each index in 0..(myOdb.Count-1)
scriptList.Add (myOdb.Get(index))
end
```

Single User Installation

Single user installations are easy. You can make your application the default project file, or ask the user to open it as a new project file. If the

204   Application Installation

user is opening the application as a new project, consider whether users should save their work over the application's original project file. In either case you can force the issue by customizing the interface. For example, if you want the user to save her work to a new file you can ask her to first perform a "save as" operation. If the following embedded script is used as the project startup script, a save as operation is performed when the user opens the project, and the script is then removed from the new project.

```
' Script name: Startup
' Get a new project name from the user.
newFileName = FileDialog.Put

("","*.apr","Save New Project As")
' Close the project if Cancel selected.
Picture frame property dialog box.
if (nil = newFileName) then
av.GetProject.Close
end

' Save the project with the new name.
thisProject = av.GetProject
thisProject.SetFileName (newFileName)
thisProject.Save
' Remove this script from the project.
thisProject.RemoveScript ("Startup")
```

To establish a system default file based on the current project, send the ***MakeSysDefault*** request to the project object. Never overwrite the system project file by copying the user project or default files to the *etc* directory. The user default file, if any, overwrites the system default file. To avoid the user default overwriting the system default file in a single user environment, set your project as the user default file so that it is always loaded. To establish the user default project file, rename the project file to *default.apr* and copy it to one of the following directories:

❏ %HOME% for MS-Windows 3.1 and NT

❏ ArcView application folder for Macintosh

❏ $HOME for UNIX

❏ SYS$LOGIN for Open VMS

The serial number of the ArcView running your application can be added to a script to provide additional security. In the following code segment, the ArcView serial number is compared to a preassigned number. If they are not the same, the application shuts down.

```
' Get ArcView's serial number.
thisSerial = av.GetSerialNumber
' Compare it to a preassigned number.
if (thisSerial whatShouldBe) then
MsgBox.Error ("This is an unauthorized copy",
"")
Av.GetProject.Close
end
```

Network Installation

Security and data access concerns in a network installation of your application are similar but more complicated than in a single user context. First and foremost, a detailed understanding of the network and the ArcView configuration is necessary.

ArcView uses networking software to access remote data. Although remote data is handled in the same fashion as local data, the drive name assigned to the remote disk must be an input into your application. Next, because ArcView does not accommodate every version of every networking software package, you should consult ArcView installation documentation and the network manager prior to a network installation.

ArcView may be installed centrally for multiple users to share, or on individual workstations. In either setup, your application's project file can be placed in a central location or distributed among workstations. Keep in mind that application maintenance is more difficult when the program is distributed to individual workstations compared to a central location.

Chapter 13

Address Matching

GIS is a tool for managing and analyzing data with spatial attributes, i.e., latitude and longitude, or northing and easting. For most people, however, street addresses are the most common spatial attribute, not X and Y. Real estate taxes are assessed by property address, crime is reported by city block, and new commercial sites are selected along major roadways.

An address specifies location in the same manner as geographic coordinates. But addresses must be converted to geographic coordinates before ArcView or any other GIS software can display and use them. The process of transforming an address (text string) into geographical coordinates is called address matching or "geocoding".

208   Address Matching

Geocoding with ArcView

ArcView provides the tools for geocoding directly through its interface or by using Avenue. Since address matching is a rather complicated process, you should familiarize yourself with address matching using ArcView's interface. Briefly, address matching with ArcView involves the following steps:

1. Make a theme matchable.

2. Add address events.

3. Create an events theme.

4. Process unmatched events.

Matched locations are depicted by graphical symbols. The address matching steps are described below.

Making a Theme Matchable

For address matching, you need a theme with specific attributes that can be searched. Each type of address format—linear, polygon and point—has different address components.

A linear address format requires a theme with line features such as a street network. Typically, the address component attributes for this format are left-from, left-to, right-from, right-to, street-name, and type. For example, an arc within a coverage could have the following values:

    left-from: 1301
    left-to: 1399
    right-from: 1300
    right-to: 1398
    street-name: WISCONSIN
    type: AV

The from-to ranges indicate the possible numbers that could fall within a specific block, and the left and right sides divide the numbers into odd and even groups. The preceding example showed values for a block along Wisconsin Avenue. Address numbers along the block range from 1300 to 1399, with odd numbers on the left and even numbers on the right. Additional optional fields such as direction can be used to further define an address.

The linear format, used most frequently for locating addresses within urban areas, can also contain zone information. Zone refers to an additional attribute on each side of the line segment in order to discriminate among streets with similar names. For example, a zip code or city name can serve as zone information. Thus, upon searching for Main Street within a specific county, a city name can narrow the search scope. ZIP codes used as zone attributes follow:

> left-zone: 20016
> right-zone: 20017

If an entire street segment is inside a ZIP code area, the left and right values in the above example would be the same. Attribute names defining address components can be different from the fields mentioned here. In the process of making a theme matchable, ArcView searches for default attribute or field names (left-from, l-f, l_f, street-name, st-name, st-n, etc.). However, if ArcView does not find what the program identifies as a particular attribute name, or selects the wrong attribute for what the user has in mind, the field names that define the address components can be set or changed.

The polygon address format requires themes with polygon features. In this format, a single attribute can be assigned as the address component, such as census tract or ZIP code area. In this case, the value of the census tract number or ZIP code is used in search and match routines.

The point address format requires themes with point features, and is useful in locating items with identifying names or numbers. Similar to the polygon format, only a single attribute is necessary. For instance, in a theme showing sewer system valves, the single attribute could be valve number. Another example is a theme comprised of landmarks where the attribute is building name.

Adding Address Events

Events are tabular data that can be added to views as themes. Initially, events are added to the project in the form of tables which contain geographic information such as street addresses or building names. Events on a base theme can be located by using such geographic information. As described later in this chapter, the base theme is created from a table. The base theme contains the address components described in the previous section.

An example of an address event is students' home street addresses in a particular school district. The base theme for this address event could be a street map of the school district. Another example of an address event is the number of home mortgages by census tract, where the base theme is a map of several contiguous census tracts.

Creating an Events Theme

Once a base theme is matchable and an appropriate events table is added to the project, ArcView attempts to locate each event on the base theme. In this process, ArcView creates a shape file theme which contains a symbol at the location of each matched event. The geocoded theme contains point features for both matched and unmatched events. The features for unmatched events are placed in the theme as null point shapes.

Processing Unmatched Events

As mentioned above, the geocoded theme created by ArcView contains both matched and unmatched events as point features. The unmatched events are represented by null point shapes in the theme. Any event in the geocoded theme can be rematched. The two approaches for processing unmatched events are manual editing to correct errors in events, and relaxation of the matching requirements to increase the probability of finding a match.

Geocoding with Avenue

Avenue can be used to install automatic address matching tasks in your application. This approach is useful when the same type of event data must be repeatedly matched to the same base theme. For example, if a monthly crime report includes a map of all incidents, you can create a script to match the crime database against your base theme. The following sections describe how to make a theme matchable and how to create an address events theme.

Avenue can also be used to add event tables. Adding event tables is similar to adding any attribute table to a project. (See Chapters 8 and 9 on adding attribute tables.)

Processing unmatched events is generally an on-line manual task because it involves reviewing why an event failed to match, and perhaps correcting data entry errors. There is little value in using Avenue for processing unmatched events unless you decide to rematch everything after relaxing match requirements.

The steps required to make a theme matchable and to create an address events theme are shown in the following two scripts. In the first script a theme called *Street* in the *County* view is made matchable for the linear address format. In the second script, an event table with addresses for bank locations is matched against the *Street* theme. The steps are explained in detail following the scripts.

```
' Prepare for address matching by making
' a theme matchable.
theProject = av.GetProject
theTheme = theProject.
FindDoc("County").FindTheme("Street")
if (nil = theTheme) then
MsgBox.Error (
"Unable to access Street theme in County view",
"")
exit
end
```

212   Address Matching

```
'
' Verify whether the theme is already matchable.
if (theTheme.isMatchable) then
MsgBox.Warning ("Street is already matchable",
"")
exit
end

'
' The GetDefStylesODB request to the AddressStyle
' class returns a file name object corresponding
' to the style object database supplied with
' ArcView.
addrStyleFilename = AddressStyle.GetDefStylesODB
if (nil = addrStyleFilename) then
MsgBox.Error (
"Unable to find the default address style ODB",
"")
exit
end

'
' The list of styles can be extracted from the
' address style file name, and the desired style
' can be found within that list.
addrStyleList = AddressStyle.GetStyles
(addrStyleFilename)
addrStyle = AddressStyle.FindStyle
("US Streets with Zone")
'
' Associate the known fields to
' the address components.
theVTab = theTheme.GetVTab
attList = {}
'
' Set up for the US Streets with Zone style.
```

Geocoding with Avenue 213

```
nameList = {"LADD_FROM", "RADD_RIGHT",
"LADD_TO", "RADD_TO", "NONE", "NONE",
"STREET", "TYPE", "NONE", "LZIP", "RZIP"}
for each fldName in nameList
if (fldName = "NONE") then
' NONE indicates that there is no field to match.
attList.Add ("") 'add a null value
continue
end

' Get the field object.
aField = theVTab.FindField (fldName)
if (nil = aField) then
MsgBox.Error ("Unable to access required field"++
fldName++,"")
exit
end

attList.Add (aField)
end
'
' Create a MatchSource object.
aMatchSource = MatchSource.Make (addrStyle,
theTheme, attList)
'
' Now assign the MatchSource object to the theme.
theTheme.Set MatchSource (aMatchSource)
'
' Verify whether theme is matchable.
if (theTheme.IsMatchable) then
MsgBox.Info (theTheme.GetName++"is matchable",
"")
else
MsgBox.Warning (theTheme.GetName++
"Is not matchable",
```

214   Address Matching

```
"")
end
' ----------------------------------------------

' Perform address matching.
'
' Get the MatchSource object.
theTheme = av.GetProject. FindDoc("County")
FindTheme("Street")
if (theTheme.IsMatchable) then
theMatchSource = theTheme.GetMatchSource
else
MsgBox.Warning (theTheme.GetName++
"Is not matchable",
"")
exit
end

'
' Get the event table VTab object.
theVTab = av.getProject.FindDoc("Banks").GetVTab
'
' Verify that MatchSource uses the zone,
' then get zone and address fields
' from the event table.
if (theMatchSource.HasZone) then
zoneField = theVTab.FindField ("ZIP")
end
addrField = theVTab.FindField ("ADDRESS")
'
' Create GeoName object.
theGeoName = GeoName.Make (theMatchSource,
theVTab, addrField, zoneField)
'
' Set a shape file name to store the new
' geocode theme.
```

Geocoding with Avenue 215

```
theGeoName.SetOutFileName ("BANKS.SHP".AsFileName)
'
' Initialize the geocode theme.
geoVTab = theMatchSource.InitGeoTheme (theGeoName)
'
' Use US_ADDR address standardization
' file for US Streets style and US Streets
' with Zone style.
aMatchKey = MatchKey.Make ("us_addr.stn")
if (nil = aMatchKey) then
MsgBox.Error ("Unable to create a match key",
"")
exit
end

'
' Create a MatchCase object.
aMatchCase = MatchCase.Make
(theMatchSource, aMatchKey)
'
' Create default matching preference values.
aMatchPref = MatchPref.Make
'
' Get number of records to match.
numRecs = theVTab.GetNumRecords
'
' Start the loop to go through all records.
for each recNum in theVTab.GetDefBitMap
' Set the progress bar.
av.SetStatus (((recNum+1)/numRecs)*100)
' Standardize the address string.
aMatchKey.SetKey (theVTab.ReturnValueString
(addrField,i))
' Search for candidates; search request returns
' number of candidates.
```

216 Address Matching

```
numCand = theMatchSource.search (aMatchKey,
aMatchPref.GetPrefVal(#MATCHPREF_SPELLWEIGHT),
aMatchCase)
if (numCand = 0) then
theMatchSource.WriteUnMatch (recNum, aMatchKey)
else
' Score and get the best candidate.
aMatchCase.ScoreCandidates
cand = aMatchCase.GetBestCand
' Write it to the GeoName data source.
TheMatchSource.WriteMatch
(recNum, aMatchKey, cand)
end
end
'
' End the matching process; Avenue closes index file.
theMatchSource.EndMatch
'
' Create and add the address event theme.
geoTheme = Theme.Make (theGeoName)
theView.AddTheme (geoTheme)
' -----------------------------------------------
```

Making a Theme Matchable

A **MatchSource** object is created in order to make a theme matchable with Avenue. The **Make** request for the **MatchSource** class requires a theme, an address style, and a list of fields in the theme's attribute table that corresponds with the fields required by the address style. Steps for this task follow:

1. Select a theme.

2. Select an address style.

3. Associate attribute fields.

4. Create a MatchSource object.

Selecting a Theme

As shown in the following code segment, a theme is selected by accessing its object. When a **MatchSource** object called *aMatchSource* is created later in the chapter, *theTheme* becomes matchable by setting its Match-Source attribute.

```
theProject = av.GetProject
theTheme = theProject.FindDoc
("County").FindTheme("Street")
if (nil = theTheme) then
MsgBox.Error
("Unable to access Street theme in County view",
"")
exit
end
'
' Check to see if theme is already matchable.
if (theTheme.isMatchable) then
MsgBox.Warning ("Street is already matchable",
"")
end
```

Selecting an Address Style

The address style defines the address format and its required and optional fields. ArcView's address styles appear below:

❏ US Streets with Zone

❏ US Streets

❏ Single Field

❏ Zip+4

218   Address Matching

The first two are linear address formats while the last two may be polygon or point address formats. Styles are stored in an object database in the *geocode* directory, and the file name for this object database is *addstyle.db*.

Creating an address style object is a two-step process:

1. Access the object database containing the address styles.

2. Search for the desired address style.

An ***AddressStyle*** class object is created when an address style is selected. In the following code segment, the default style database supplied with ArcView is accessed, and the *US Streets with Zone* style is located.

```
' GetDefStylesODB request to the AddressStyle class
' returns a file name object corresponding to the
' style object database supplied with ArcView.
addrStyleFilename = AddressStyle.GetDefStylesODB
if (nil = addrStyleFilename) then

MsgBox.Error ("Unable to find the default address
style ODB",
"")
exit
end

'
' The list of styles can be extracted from the
' address style file name and the desired style
' can be found in the list.
addrStyleList = AddressStyle.GetStyles
(addrStyleFilename)
addrStyle = AddressStyle.FindStyle ("US Streets with
Zone")
```

Each style has a set of default parameters that can be reviewed and altered through Avenue requests. For example, when an event is located, a symbol is placed at a predefined distance from the street. This distance, measured in map units, is called the "offset value". As shown in the following statements, the default offset value for an address style can be changed.

```
' Change offset value and save it to a
' new style object database for future use.
addrStyle.SetDefOffset (newValue)
AddressStyle.SaveStyles ("mystyle.db".AsFileName)
```

Associating Attribute Fields

An address style must be selected prior to associating fields from the event table. An address style maintains the required and optional address components in a predetermined order. To associate attribute fields from the event table to address components, an event table's list of fields must be created in the same order. If optional address components exist, and you do not wish to provide corresponding attribute fields, nil objects corresponding to these components are stored in their respective places on the list.

Creating the fields list can be tricky. Unless the field names are known ahead of time, the user may be requested to select the fields. For example, if the same event table is matched every month, you could code the field names directly into your script. However, if the event table and its fields are not always the same, you should display the fields list and ask the user to determine how the fields correspond to the address components. Both approaches are presented in the following two code segments. In the first segment, the field names are known ahead of time. In the second, a complete fields list is presented to the user for selecting each component.

```
' Associate the known fields to the address
' components.
theVTab = theTheme.GetFTab
attList = {}
'
' Set up for the US Streets with Zone style.
nameList = {"LADD_FROM", "RADD_RIGHT",
```

220 Address Matching

```
"LADD_TO", "RADD_TO", "NONE", "NONE",
"STREET", "TYPE", "NONE", "LZIP", "RZIP"}
for each fldName in nameList
if (fldName = "NONE") then
' NONE indicates that we have no field to match.
attList.Add ("") ' Add a null value.
continue
end

' Get the field object.
aField = theVTab.FindField (fldName)
if (nil = aField) then
MsgBox.Error ("Unable to access required field"++
fldName++,"")
exit
end

attList.Add (aField)
end
' ---------------------------------------

' Associate the event table fields with
' address components by presenting them
' to the user; store the results in attList.
theVTab = theTheme.GetFTab
attList = {}
'
' Get the list of address components.
matchFieldList = addrStyle.GetMatchFields
'
' As each address component is presented
' to the user, identify it as required or optional.
for each fld in MatchFieldList
if (fld.IsRequired) then
optionality = "Required"
else
```

Making a Theme Matchable 221

```
optionality = "Optional"
end
while (True) ' Stay in loop until you get what
' you want.
aField = MsgBox.List (theVTab.GetFieldList,
fld.GetName++optionality,"")
' aField is nil if user clicked on the Cancel button.
if (nil = aField AND optionality = "Required") then
' Null values cannot be allowed for required
' components.
stopIt=MsgBox.YesNo
("Do you want to cancel address matching?",
"", False)
if (stopIt) then
exit
else
' Try again to get the field name.
continue
end
end
break
end
' Add field name to the list of attribute fields.
attList.Add (aField)
end
' ----------------------------------------
```

The Geographic Dimension of Data

By MatchWare Technologies

The population, scope, and variety of "corporate databases" currently in operation, design and development are limitless. These databases reside on mainframes, minicomputers and desktop computers across the private and public sectors, and are accessed, queried, and updated millions of times per day in order to answer a variety of questions dealing with product and/or customer status. However, few of these databases can be readily accessed based on geographical parameters such as "How many of these products have we sold within 100 miles of X distribution center?" or "How many customers do we have within a two-hour drive of the Y trade show location?"

Several database management systems currently offer geographic query options, although their popularity seems to be developing rather slowly. Next, it is interesting that the vast majority of records in today's corporate databases contain *geographic codes* of one kind or another which would support diverse geographic analysis applications. Geocodes include postal codes (five-digit zip codes, nine-digit zip codes, and carrier route codes); census codes (MCD, MSA, places, tracts, block groups, and blocks); latitude and longitude coordinate codes; and plain vanilla addresses of consumers and businesses. Geocoding is the matching of geographic codes in records with geographic directories in order to determine common geographic overlap.

As spatial relationships of customer locations, sales coverage, stores, and distribution centers continue to emerge as major factors in critical decision-making, the integration of geography as a third dimension of database access and analysis methodology will rapidly expand in popularity and use. This trend will be accelerated by the continued expansion of desktop computing capabilities and the rapid improvement of desktop price-to-performance ratios. Even though desktop computers are not particularly well-suited for cartographic technology, they are superb spatial data analysis tools.

It is noteworthy that many spatial data analysis applications have been in use for over two decades. For instance, in the early 1970s, R.L. Polk and Donnelley Marketing provided census tract coding services to support market segmentation and profiling, and Urban Data Processing (then known as Harte Hanks Data Technology) offered Area Profile Reports which aggregated census demographic data around retail store locations. Both companies' services, however, would have been impossible without the availability of computerized geographic databases developed for the public domain by the U.S. Census Bureau.

The following short list of major projects using address matching demonstrates the diverse nature and utility of geographic spatial analysis at present and in the future.

❐ The U.S. Department of Transportation requires that all pipeline operators shall "establish a continuing education program to develop public awareness and ability to recognize hazardous pipeline emergencies and report same. This program is to be implemented for the population within the coverage of the pipeline's operation." In order to use the capability of direct mail to efficiently target specific controlled audiences, the Harte Hanks Direct Marketing division adopted GIS methodology to fulfill and manage compliance with the D.O.T. regulation. Harte Hanks used latitude/longitude records of pipeline paths to define "corridor" buffer zones within a quarter-mile of pipeline positions. After the buffer zone files were complete, they were matched against compiled mailing lists which were also coordinate coded. The application of point-in-polygon algorithms made possible identification of addresses which fell inside the buffer zone and thus, who should participate in the program.

❐ A major overnight express delivery service offers a variety of service levels including regular daily pick-up, on-call request pick-up, and drop-off boxes in convenient locations. Regular pick-up service reflects high volume clients, while customers who use on-call pick-up incur a billable charge. The company decided to encourage increased use of the drop-off boxes in hopes of increasing efficiency, particularly by on-call customers because the pick-up charge did not cover costs. In addition to elimination of pick-up fees for clients, the use of drop-off boxes offered the promise of improved service. However, the use of drop-off boxes remained at a low level several months after the company adopted the policy for enhanced efficiency.

The company contracted Geographic Data Technology (GDT) in Lyme, New Hampshire, to geocode customer addresses and drop-off box addresses with latitude/longitude coordinate codes. Although the match rate was not 100%, the less than perfect rate was anticipated. The match rate was judged to be more than satisfactory for purposes of the program. GDT was able to use the coordinate coded files to relate customer locations to the nearest drop-off box in most cases, and a proximity index was established. After the proximity index was completed, most customer records were modified to reflect the location of the nearest drop-box.

When a customer called in and requested pick-up service, the operator checked the customer's screen record to determine whether a convenient drop-box was located nearby. If a drop-box location appeared convenient for the caller, the operator would encourage its use as an alternative to pick-up service. Results of this GIS application were positive: on-call pick-up service requests declined; customer satisfaction remained unchanged; customer costs decreased; and market area revenue increased.

❐ When management of the largest pizza chain in the United States decided to offer home delivery service, the company was faced with several daunting challenges, not the least of which was a well-positioned competitor. One significant advantage enjoyed by the dominant supplier was that consumers often knew which store to call for service in their area. Key tactical objectives established by the new entrant included the ability to mass merchandise the delivery service in support of all participating stores within each market, and to match the delivery time characteristics to which the market had become accustomed.

Management decided to offer a single telephone order entry number for an entire market area, and operators were instructed to direct the orders to appropriate store locations for preparation and delivery. To maximize efficiency, it became necessary to define and maintain delivery market

> zone definitions for each store location and immediately match the address of a caller to a delivery zone, and thus, to assign the order to the zone's store. If matches could not be established, it seemed appropriate to show appreciation to such callers, and to collect their addresses for use in future market studies.

Creating a MatchSource Object

A *MatchSource* object is a feature source that accepts matching requests. When a MatchSource is created, a theme is made matchable. Creating a *MatchSource* object is easy once the theme, address style, and fields list are selected or prepared. These objects are simply supplied as parameters of the *Make* request to the *MatchSource* class. See the following code segment for the format of the Make request.

```
' Create a MatchSource object.
aMatchSource = MatchSource.Make (addrStyle,
theTheme, attList)
'
' Now assign the MatchSource object to the theme.
theTheme.SetMatchSource (aMatchSource)
'
' Check to verify that the theme is matchable.
if (theTheme.IsMatchable) then
MsgBox.Info (theTheme.GetName++"is matchable",
"")
else
MsgBox.Warning (theTheme.GetName++
"Is not matchable","")
end
```

226   Address Matching

Creating an Address Events Theme

The process of address matching results in a new theme that displays a symbol for each matched event. The following steps are required for address matching:

1. Retrieve the match source.

2. Select the event table.

3. Initialize a geocode theme.

4. Match addresses.

5. Create an address event theme.

Retrieving the Match Source

A matchable theme has a ***MatchSource*** object. This object is necessary in order to search for each address event. In the following script, a theme is tested to determine whether it is matchable, and then the MatchSource object is retrieved.

```
' Get the MatchSource object.
theTheme = av.GetProject. FindDoc("County")
FindTheme("Street")
if (theTheme.IsMatchable) then
theMatchSource = theTheme.GetMatchSource
else
MsgBox.Warning (theTheme.GetName++"Is not matchable",
"")
end
```

Selecting the Event Table

The event table holds the field or fields which describe the address. First, the event table and the fields to match must be accessed. Given the use of the US Streets with Zone address style, two fields from the event table are required. One field holds the address while the other field contains the zone.

The contents of the zone field or the address field for the polygon or point address format are simple. In these cases, a single value is matched against another single value in the matchable theme. The address field for the linear address style requires a more complicated procedure because the field is composed of several components, and each is matched against a different field in the matchable theme. Because the address field in the event table is a single text field, ArcView must follow a set of rules to identify each address component in the field. For example, ArcView "assumes" that the first numerical value is the house number, and that a single alphanumeric (e.g., *W*) before or after the street name indicates direction (west). Standards for identifying address components are found in the ArcView documentation.

The following code segment shows how to create address and zone field objects.

```
' Retrieve the VTab object for the event table.
theView = av.GetProject.FindDoc ("County")
theTheme = theView.FindTheme ("Street")
theMatchSource = theTheme.GetMatchSource
theVTab = theEventTable.GetVTab
'
' Verify that MatchSource uses zone;
' then retrieve zone and address fields
' from the event table.
if (theMatchSource.HasZone) then
zoneField = theVTab.FindField ("ZIP")
end
addrField = theVTab.FindField ("ADDRESS")
```

Initializing a Geocode Theme

A geocode theme is a feature source that holds a record for each record in the event table. Initially, a geocode has a null shape for each event and its match status is set to *unmatch*. The new theme is geocoded with the **Search** or **Rematch** requests, as discussed in the next section.

A **GeoName** object, which contains the information required to create a geocoded data source, is necessary in order to initialize a geocode theme. This object is created from the *MatchSource*, event table, address field, and zone field. In addition, an output file name must be assigned to store the geocoded data set in a shape file format. In the following code segment, a GeoName object is created, and then used to initialize a geocode theme.

```
' Create GeoName object.
theGeoName = GeoName.Make (theMatchSource,
theVTab, addrField, zoneField)
'
' Set a shape file name to store the new
' geocode theme.
theGeoName.SetOutFileName ("BANKS.SHP".AsFileName)
'
' Initialize the geocode theme.
theMatchSource.InitGeoTheme (theGeoName)
```

Matching Addresses

Search and **WriteMatch** requests to the **MatchSource** object are used to find a match and write it to the geocode theme. Before using these requests, however, three objects from the **MatchKey**, **MatchCase**, and **MatchPref** classes must be created.

The **MatchKey** class standardizes the address field of the event table. A **MatchKey** object is created from the information supplied in standardization files. ArcView provides standardization files for each address style. These files have an *stn* extension, and reside in the *geocode* directory. The standardization files are related to several other files which define expected address components, possible and allowed

Creating an Address Events Theme   229

values, and a scoring system. For example, for the US Streets address style defined in the *us_addr.stn* file, Avenue searches for the following components in an address field:

- ❏ House number
- ❏ Pre-direction
- ❏ Pre-type
- ❏ Street name
- ❏ Suffix type
- ❏ Suffix direction
- ❏ Soundex of street name
- ❏ Reverse soundex of street name

While a "hit" for every component is not expected, the higher the number of matched components, the higher the match score. A match score is computed to assist in selecting the best among multiple matches. The following code segment creates a **MatchKey** object for the US Streets address style.

```
' Use US_ADDR address standardization
' file for US Streets style and US Streets
' with Zone style.
aMatchKey = MatchKey.Make ("us_addr.stn")
if (nil = aMatchKey) then
MsgBox.Error ("Unable to create a match key",
"")
exit
end
```

A **MatchCase** object is comprised of a list of matching candidate records for a single event. The candidates can be scored to select the best match. In the following code line, a **MatchCase** object is created. This object is populated each time a **Search** request is sent to the **MatchSource** object.

```
' Create a MatchCase object.
aMatchCase = MatchCase.Make (theMatchSource,
aMatchKey)
```

230 Address Matching

A ***MatchPref*** class object contains geocoding preferences such as the minimum match score. Default preference values are stored in the *mprefdef.db* file in the *geocode* directory. Default preference values are listed below. (See the ArcView documentation for a discussion of these parameters.)

❏ Spell weight: 80

❏ Minimum match score: 50

❏ Minimum candidate: 30

❏ No user review: 1

Through the ***MatchPref*** class, Avenue allows you to change the default values to match your needs. A MatchPref object is created in the code line below. Because the attributes for this object are not changed, the default values are assumed.

```
' Create default matching preference values.
aMatchPref = MatchPref.Make
```

Once the ***MatchKey***, ***MatchCase***, and ***MatchPref*** objects are created, address matching can begin. This process generally requires iteration through all records of the event table in an attempt to match each record. The matching process for each event involves the following tasks:

1. Standardize the address string.

2. Send a ***Search*** request to the ***MatchSource*** object.

3. If no matching candidate is found, send a ***WriteUnMatch*** request to the ***MatchSource*** object.

4. If matching candidates are found, score them, select the best candidate, and send a ***WriteMatch*** request to the ***MatchSource*** object.

The preceding tasks are demonstrated in the following code segment. A progress bar showing the proportion completed as matching occurs is included.

Creating an Address Events Theme   231

```
' Get number of records to match.
numRecs = theVTab.GetNumRecords
'
' Start the loop to go through all records.
for each recNum in theVTab.GetDefBitMap
' Set the progress bar.
av.SetStatus (((recNum+1)/numRecs)*100)
' Standardize the address string.
aMatchKey.SetKey (theVTab.ReturnValueString
(addrField,i))
' Search for candidates; search request returns
' number of candidates.
numCand = theMatchSource.search (aMatchKey,
aMatchPref.GetPrefVal(#MATCHPREF_SPELLWEIGHT),
aMatchCase)
if (numCand = 0) then
theMatchSource.WriteUnMatch (recNum, aMatchKey)
else
' Score and retrieve the best candidate.
aMatchCase.ScoreCandidates
cand = aMatchCase.GetBestCand
' Write the best candidate to the
' GeoName data source.
TheMatchSource.WriteMatch (recNum, aMatchKey, cand)
end
end

'
' End the matching process; Avenue closes index file.
theMatchSource.EndMatch
```

232   Address Matching

Creating the Address Event Theme

After creating the *GeoName* data source, initializing the address theme, and recording the match results, the only remaining task is to create the theme. In the following code segment, the address event theme is created and added to the view.

```
' Create and add the address event theme.
geoTheme = Theme.Make (theGeoName)
theView.AddTheme (geoTheme)
```

Chapter 14

Integration

The ability to access system resources enhances an application's capabilities. Avenue allows you to access the system clipboard and environment variables, and to issue operating system commands inside your program. Avenue's integration capabilities, however, go far beyond these tasks.

Avenue supports Dynamic Data Exchange (DDE) and Remote Procedure Call (RPC). For data transfer between applications, DDE is used in the Microsoft Windows environment, and RPC in the UNIX environment. For example, if you use a spreadsheet which supports DDE and you have created a budget analysis model using numbers from ArcView, the spreadsheet can be linked to ArcView. As the numbers change in ArcView, spreadsheet numbers are automatically updated.

This chapter covers accessing the clipboard and system variables, issuing system commands, and the use of DDE and RPC. As an extension of issuing system commands, the execution of ARC/INFO AMLs is also discussed. Topics discussed in this chapter are generally platform dependent. Consequently, attention to the specific class or request for your hardware and operating system platform is required.

234   Integration

Accessing the Clipboard

The system clipboard holds text strings and graphics, while the application (ArcView) clipboard holds ArcView objects. The graphics in the system clipboard are in metafile format for Windows, and PICT format for Macintosh.

The **Clipboard** class in Avenue references the application clipboard. Although ArcView does not have a class for the system clipboard, text strings can be exchanged with the system clipboard through the application clipboard. Graphics cannot be moved between the two, and the two clipboards do not automatically synchronize.

The following code line shows how the clipboard object is accessed.

```
appClipBoard = ClipBoard.The
```

Because the clipboard object is a specialized list, you can use the **Add** and **Get** requests to copy an object to the application clipboard or paste from it. The **Add** request clones an object and adds the clone to the clipboard. As seen in the following code segment, the object is cloned in order to paste from the clipboard.

```
' Place the first two themes in View1
' on the clipboard.
appClipBoard = ClipBoard.The
theView = Av.GetProject.FindDoc ("View1")
appClipBoard.Add (theView.GetThemes.Get(0))
appClipBoard.Add (theView.GetThemes.Get(1))
'
' Paste the themes on the clipboard to
' the View2 document.
newView = Av.GetProject.FindDoc ("View2")
newView.AddTheme (appClipBoard.Get(0).Clone)
newView.AddTheme (appClipBoard.Get(1).Clone)
```

To remove all objects from the application clipboard, send the **Empty** request as shown in the code line below.

```
ClipBoard.The.Empty
```

Accessing the Clipboard 235

Text can be moved between the system and application clipboards. When you send the ***Update*** request to the ***ClipBoard*** class object, ArcView checks to see if there are any objects in the application clipboard. If one or more objects are found, the contents of the application clipboard are copied to the system clipboard. In the event no objects are found in the application clipboard, ArcView then checks for a new item in the system clipboard. If there is a new item in the system clipboard it is added to the application clipboard.

The string objects in the application clipboard are converted to text blocks before copying to the system clipboard; and text in the system clipboard is converted to a string object when copied to the application clipboard. Graphics and objects of other classes are ignored and not transferred. This entire transfer process occurs when the ***Update*** request is sent to the clipboard object. A Boolean object is returned by the ***Update*** request, and is set to True if the system clipboard had a new item since the last update. The following code segment shows how the ***Update*** request works.

```
' Synchronize the clipboards.
appClipBoard = ClipBoard.The
newClip = appClipBoard.Update
if (newClip) then
' The contents of the system clipboard are
' added to the application clipboard.
MsgBox.ListAsString (appClipBoard,
"Contents of the application clipboard",
"")
else
if (appClipBoard.Count  0) then
' The contents of application clipboard
' are copied to the system clipboard.
MsgBox.ListAsString (appClipBoard,
"Contents of the application clipboard",
"")
```

236 **Integration**

```
else
MsgBox.Info ("Clipboards are empty", "")
end
end
```

Accessing the Operating System

The *System* class provides the tool for accessing operating system variables and commands. For example, if you want to start another program or obtain the value of an environment variable, appropriate requests are sent to the System class. The System class has no objects; all requests are sent directly to the class.

Accessing Environment Variables

Environment variables are retrieved or set through the *GetEnvVar* and *SetEnvVar* requests. For example, to locate the directory that holds ArcView data, the *AVDATA* value can be retrieved, assuming this variable is set in the *autoexec.bat* file. The following code segment searches for *AVDATA*; if *AVDATA* is not found, the variable is set to a user-supplied value.

```
' Get the AVDATA environment variable.
avData = System.GetEnvVar ("AVDATA")
if (nil = avData) then
' If not set, get AVDATA value from the user, and
' then set it.
avData = MsgBox.Input ("AVDATA variable
is not set",
"Enter AVDATA value: ","")
if (nil = avData) then
exit
end
System.SetEnvVar ("AVDATA",avData)
end
```

Additional valuable information can be obtained from the system. For example, screen resolution in pixels can be determined by sending the **ReturnScreenSizePixels** request. This request returns a **Point** object. The object's X value is screen width, and the Y value is screen height.

Issuing Operating System Commands

The **System** class can be used for direct issue of system commands. For instance, you can easily create a new directory, or copy one file to another by using system commands. As shown in the following code segment, the **Execute** request can issue system commands. The contents of the command line (*commandLine*) are platform dependent.

```
' Make a quick backup of the project file.
projectName = Av.GetProject.GetFileName
baseName = projectName.GetBaseName
backupName = baseName.SetExtension ("BAK")
commandLine =
"Copy"++projectName.AsString++backupName
System.Execute (commandLine)
```

Executing AMLs

AMLs are executed from an Avenue script by sending the **Execute** request to the **System** class. The code below runs an AML called "getvalve".

```
' Run AML buffer.
System.Execute ("arc getvalve")
```

Avenue does not wait for the **Execute** request to complete its process. As soon as the command line is passed to the operating system, Avenue executes the next line of the script.

Implementing Dynamic Data Exchange

Dynamic Data Exchange (DDE) in Microsoft Windows is used to create a relationship between data stored in one application with data stored in another. This mechanism enables two applications to continuously and automatically exchange data. Both applications must support DDE to be able to participate in this data exchange.

In a DDE conversation, the application that initiates the conversation is known as the "client", and the responding application is called the "server". ArcView can be both client and server simultaneously.

To initiate a DDE conversation and exchange data, the client application must specify the following:

❐ Server application name

❐ Conversation topic

❐ Item

The name of the server application and its executable file are often the same. For instance, if you have created a Visual Basic program named *myvb.exe*, its server name is *myvb*. The server name of commercial applications that support DDE is included in respective documentation. An example here is the server name *WinWord* specified in the Microsoft Word for Windows documentation.

The conversation topic is usually a data unit meaningful to the server. For example, in Microsoft Word, the topic can be a file name. A DDE conversation begins after the server recognizes the topic. Once a DDE conversation is established, the application or topic name cannot be changed. Many applications that support DDE also support a topic called *System*.

Finally, "item" is the term for the piece of data to be exchanged, such as a field value in a database application.

DDE conversations are often called "hot links" or "cold links". The difference between the two types of link is in how server data is updated when changes occur. In a hot link, the server supplies new data to the client every time the data changes. In a cold link, the client must request the new data.

ArcView as DDE Client

Through Avenue, ArcView can initiate a DDE conversation. To begin a DDE and receive data, you need to know the server, topic and item names. Because all three are application dependent, a review of the appropriate documentation is recommended.

For example, if the server application is a Visual Basic program, the executable file name minus its extension is the server name. The topic name is set for each Visual Basic form as its *LinkTopic* property and it can be any character string. The *Name* property of a control on the Visual Basic form becomes the item name.

Let's assume that you have a Visual Basic program named *myvb.exe* which contains a form. The form's LinkTopic property is set to *Form1*. The form has a text box control named *Text1*. In the following code segment, a DDE conversation is established with *myvb*, and the contents of its text box are retrieved. Keep in mind that *myvb* must be running in order to establish a DDE conversation. The program could be minimized on your screen.

```
' Establish a DDE conversation with myvb.exe.
aDDEClient = DDEClient.Make ("myvb","Form1")
if (aDDEClient.HasError) then
MsgBox.Error (aDDEClient.GetErrorMsg, "")
exit
end
'
' Get the text string in the Text1 text box.
textValue = aDDEClient.Request ("Text1")
```

The **Make** request sent to the **DDEClient** class requires two parameters. The first parameter is the server name and the second parameter is the topic name. Once a DDEClient object is created, you can request information, send information, or execute a task. Several DDE conversations can be opened simultaneously to the same or different servers.

240 Integration

ArcView as DDE Server

The default project startup script establishes ArcView as a DDE server. In the event the default script is changed or the DDE server is halted, the DDE server routine can be started by the following statement:

```
DDEServer.Start
```

If the above statement is executed more than once, two instances of ArcView server are established. This situation causes errors when other applications initiate DDE conversations. To avoid executing the **Start** request more than once, insert a **Stop** request before the **Start** request as shown in the following code segment.

```
' Start DDE server support, but make
' sure that it is not duplicated.
DDEServer.Stop
DDEServer.Start
```

The ArcView server name is *ARCVIEW*, and the ArcView server supports only one topic called *SYSTEM*. Items in the ArcView server can be any string type object. For example, if you are developing a Visual Basic program to show the name of the ArcView project in a text box, the text box must have the following properties:

```
LinkTopic = ARCVIEW|SYSTEM
LinkItem = av.GetProject.AsString
```

Implementing Remote Procedure Call

The Remote Procedure Call (RPC) is a logical client/server communication system that supports network applications in a UNIX environment. A client process makes an RPC and waits for a response to be returned from the server process. Developers of distributed applications avoid the details of the network interface with the use of RPC. The following information items are required to create an RPC:

1. Server machine name. This is the host name for the RPC server.

Implementing Remote Procedure Call   241

2. Server identification number. This number identifies the RPC server procedure to be called.

3. The RPC protocol version number.

ArcView can be an RPC server while simultaneously making RPC requests. The server machine name is the host name. With the use of the host name, the RPC servers running on the host can be identified. In a UNIX system, the *rpcinfo* utility provides RPC information. In the following example, the registered RPC services on a machine named *washington* are displayed:

```
/usr/etc/rpcinfo -p washington
```

The *rpcinfo* utility returns the server identification number under the program column along with the RPC version number for the program.

ArcView as RPC Client

ArcView as the RPC client can connect to an RPC server and request services. When making an RPC request, ArcView waits for the results before proceeding. If the time-out expires before the RPC server completes the request, the RPC is canceled. The default time-out is 20 seconds, but the default value can be changed. In the following example, an RPC client is established, a higher time-out value is set, and an RPC request is made.

```
' Connect to server on host named washington,
' with program ID of 0x4000001 and version number 1.
if (RPCClient.IsMachine("washington")) then
if (RPCClient.HasServer ("washington", 0x40000001,
1))
then
anRPCClient = RPCClient.Make
("washington", 0x40000001, 1)
else
MsgBox.Error
("Invalid server ID for host washington",
"")
```

242   Integration

```
exit
end

else
MsgBox.Error
("Unable to connect to host washington",
"")
exit
end

'
' Verify that connection was
' made and client is valid.
if (anRPCClient.HasError) then
MsgBox.Error (anRPCClient.GetErrorID.AsString ++
anRPCClient.GetErrorMsg, "")
exit
else
' Set the time-out to at least 25 seconds.
if (anRPCClient.GetTimeout  25) then
anRPCClient.SetTimeout (25)
end
end

'
' Make a request and store the results in
' the respRPC string object.
respRPC = anRPCClient.Execute
(1, aCommandString, String)
```

An **RPCClient** object is always created when a **Make** request is sent. Therefore, the only way to verify the success of an RPCClient is to send the **HasError** request. Next, you can verify that the server machine is on the network and supports the server identification number. Use the **IsMachine** and **HasServer** requests to the RPCClient class to verify that the server is available before attempting to make an RPCClient object.

The **_Make_** request requires three parameters: host name, server ID, and version number. The **_Execute_** request sends a command to the RPC server. The first parameter for the Execute request is a procedure ID significant to the server which identifies the function expected from the server.

ArcView as RPC Server

ArcView can also act as an RPC server. When performing as an RPC server, ArcView accepts only one type of request: script execution. The procedure ID for executing a script is one (1).

The command line is the text of the Avenue script that is executed. For example, _av.GetProject.GetName_ can be a command line. Predefined scripts can be run in the server by a command line such as _av.Run(scriptName, nil)_. ArcView always returns a string object to the RPCClient.

Only one instance of ArcView as RPC server is viable. Use the following code segment to start an RPC server.

```
' Start an RPC server, but verify
' that it is not currently running.
RPCServer.Stop
RPCServer.Start (0x4000001, 1)
```

The Solaris (SunOS 5.3) operating system recommends using server ID numbers between 0x40000001 and 0x5fffffff for customer-written applications. Verify whether server ID number ranges are recommended for other platforms.

An ArcView RPC client would use the following code segment to get the project name loaded into the ArcView server:

```
anRPCClient = RPCClient.Make
("washington", 0x40000001, 1)
anRPCClient.Execute (1,
"av.GetProject.GetName", String)
```

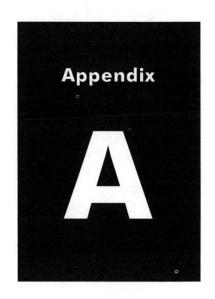

Avenue Class Hierarchy

Avenue Class Hierarchy

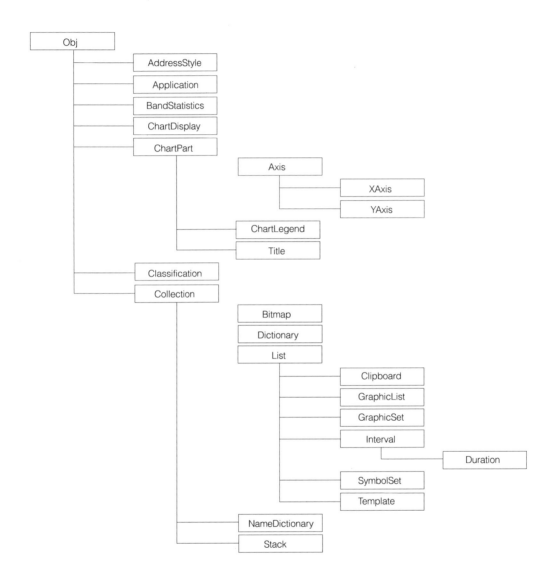

Avenue Class Hierarchy

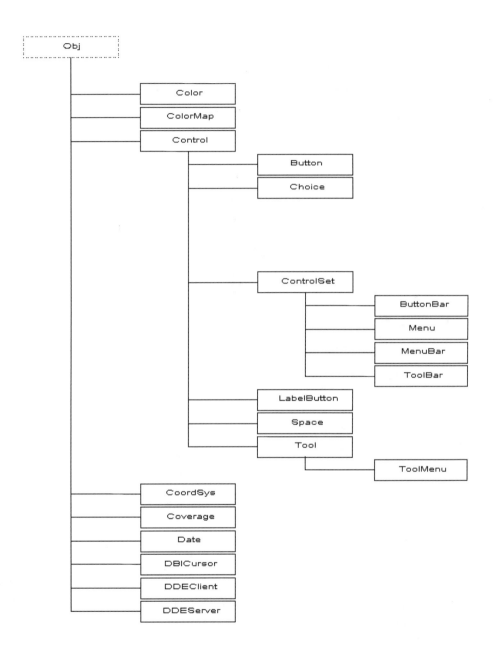

248   Avenue Class Hierarchy

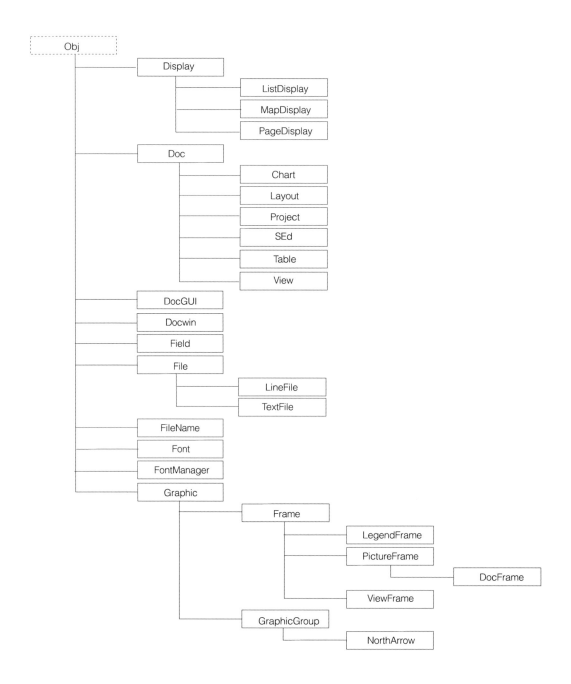

Avenue Class Hierarchy

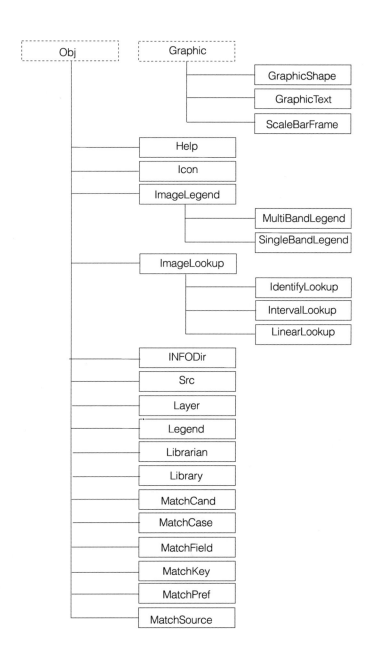

250   Avenue Class Hierarchy

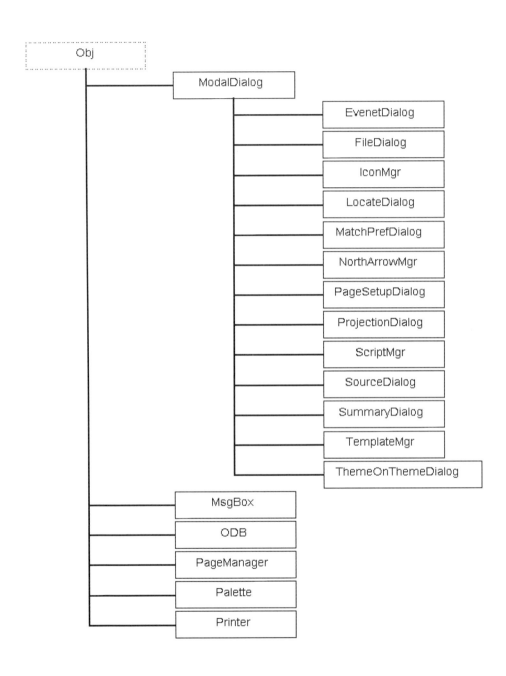

Avenue Class Hierarchy 251

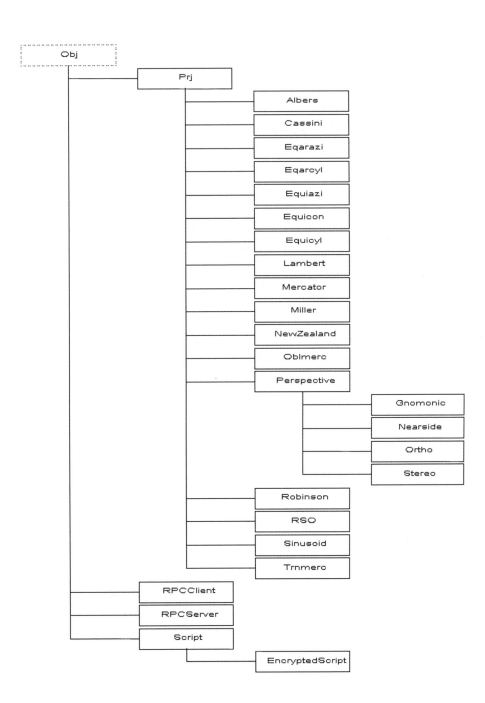

252 Avenue Class Hierarchy

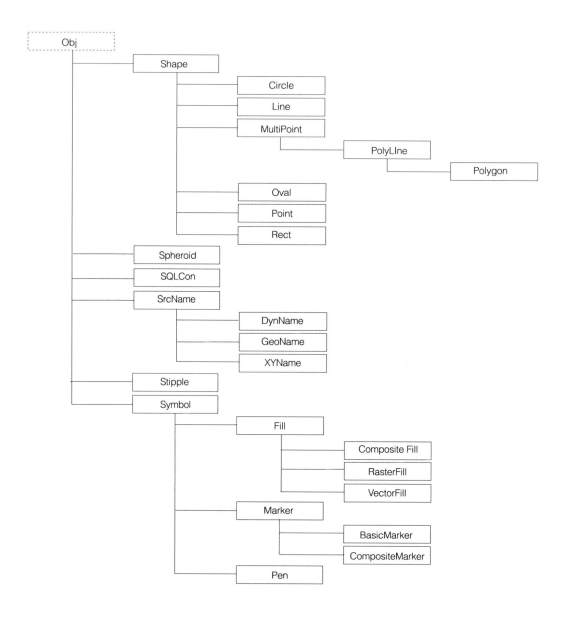

Avenue Class Hierarchy 253

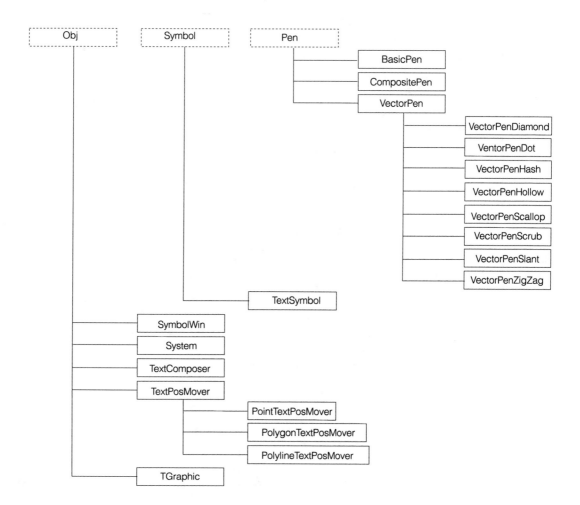

254   Avenue Class Hierarchy

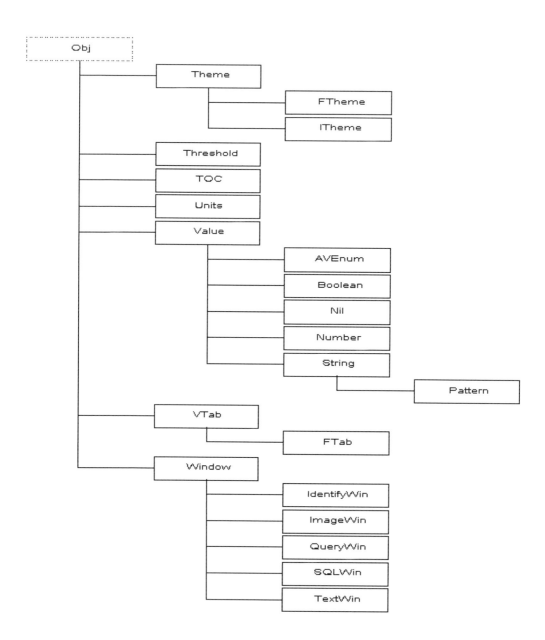

Appendix B

Avenue Reserved Words

The following words and expressions are reserved for use by Avenue, and should not be used as variable names.

256   Avenue Reserved Words

AVEnum
AddressStyle
Albers
And
Application
Axis
BandStatistics
BasicMarker
BasicPen
BitMap
Boolean
Break
Button
ButtonBar
Cassini
Chart
ChartDisplay
ChartLegend
ChartPart
Choice
Circle
Classification
Clipboard
COLlection
COLor
COLorMap
CompositeFill
CompositeMarker
CompositePen
Continue
Control
ControlSet
CoordSys
Coverage
DBICursor
DDEClient
DDEServer
Date
Dictionary
Display
Doc
DocFrame

DocGUI
DocWin
Duration
DynName
Each
Else
Elseif
EncryptedScript
End
Eqarazi
Eqarcyl
Equiazi
Equicon
EventDialog
Exit
False
FTab
FTheme
Field
File
FileDialog
FileName
Fill
Font
FontManager
For
Frame
GeoName
Gnomonic
Graphic
GraphicGroup
GraphicList
GraphicSet
GraphicShape
GraphicText
Help
INFODir
ISrc
ITheme
Icon
IconMgr
IdentifyWin

IdentityLookup
If
ImageLegend
ImageLookup
ImageWin
Interval
IntervalLookup
LabelButton
Lambert
Layer
Layout
Legend
LegendFrame
Librarian
Library
Line
LineFile
LinearLookup
List
ListDisplay
LocateDialog
MapDisplay
Marker
MatchCand
MatchCase
MatchField
MatchKey
MatchPref
MatchPrefDialog
MatchSource
Menu
MenuBar
Mercator
Miller
ModalDialog
MsgBox
MultiBandLegend
MultiPoint
NameDictionary
Nearside
NewZealand
Nil

Avenue Reserved Words 257

NorthArrow
NorthArrowMgr
Not
Number
ODB
Obj
Oblmerc
Or
Ortho
Oval
PageDisplay
PageManager
PageSetupDialog
Palette
Pattern
Pen
Perspective
PictureFrame
Point
PointTextPositioner
PolyLine
PolyLineTextPositioner
Polygon
PolygonTextPositioner
Printer
Prj
Project
ProjectionDialog
QueryWin
RPCClient
RPCServer
RSO
RasterFill
Rect
Return
Robinson
Sed
SQLCon
SQLWin
ScaleBarFrame
Script
ScriptMgr

Shape
SingleBandLegend
Sinusoid
SourceDialog
Space
Spheroid
SrcName
Stack
Stereo
Stipple
String
SummaryDialog
Symbol
SymbolSet
SymbolWin
System
TGraphic
TOC
Table
Template
TemplateMgr
TextComposer
TextFile
TextPositioner
TextSymbol
TextWin
Theme
ThemeOnThemeDialog
Then
Threshold
Title
Tool
ToolBar
ToolMenu
Trnmerc
True
Units
VTab
Value
VectorFill
VectorPen
VectorPenDiamond

VectorPenDot
VectorPenHash
VectorPenScallop
VectorPenScrub
VectorPenSlant
VectorPenZigZag
View
ViewFrame
While
Window
XAxis
XYName
YAxisY

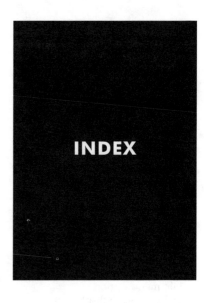

INDEX

A

Access keys, interface customization 19
Accessing files
    Close request 70
    Closing files 69
    Creating or opening files 69
    Delete request 70
    ElementCount 71
    Exists request 70
    File name object 68
    GetPos request 72
    GetSize request 72
    Make request 68, 69
    Read request 71
    Reading a file 71
    SetPos request 71
    Write request 71
    Writing a file 71
Accessing system scripts 47
AChartLoc 169
Active objects, checking for 63
Active themes, setting display extent 110
Add request 203, 234
AddBatch request 188
Adding themes 119–125

AddRecord request 153
Address events, adding 210
Address events theme creation 232
    Event table selection 227
    Geocode theme, initializing 228
    MatchSource object retrieval 226
    Zone field objects 227
Address matching
    Address events theme creation 226–232
    Attribute fields, associating 219
    Default preference values 230
    Event table selection 227
    Geocode theme, initializing 228
    Geocoding 207–232
    Geocoding with Avenue 211
    Geographic dimension of data 222
    MatchCase object 228–230
    MatchKey object 228–230
    MatchPref object 228, 230
    MatchSource object creation 225
    MatchSource object retrieval 226
    MatchSource request 230
    Search request 229, 230
    Style selection 217
    Theme selection 217

260 Index

Theme, making matchable 216–225
WriteMatch request 230
WriteUnMatch request 230
AddressStyle object 218
AGroupLabel object 168
AGroupNumber object 168
AML
ARC/INFO, applications 6
Avenue distinguished from 38
Avenue's integration capabilities 233
Executing 237
Ampersand, access key assignment 19
Application creation
Avenue script creation 81
Comments 83
Creator field 83
Customize dialog box 81
Document windows, arranging 84
Document, finding 84
File menu option 82
Help message display 85
Icons, arranging 84
Make Default button 81
Menu item, creating 81
MsgBox request 86
Naming project 82
Open Project menu 82
Project parameters, setting 83
Project Properties dialog box 83
SetStatus request 86
ShowMsg request 86
Startup script 82
Status bar display 86
Status line messages 86
Tutorial 88–101
Tutorial, application documents 89
Tutorial, Avenue scripts 95
Tutorial, development overview 88
Tutorial, Guide New Layout script 101
Tutorial, Guide Print script 100
Tutorial, Guide Select States script 96
Tutorial, Guide Update Chart script 98
Tutorial, Guide View Chart script 99
Tutorial, interface customization 90
Tutorial, menu items 88
Tutorial, testing application 102
Application files, accessing 67–71
Application overview, programming view
    documents 105

Apply events 61
Arc, SrcName object 122
ARC/INFO, applications 6
ArcStorm
Data source, accessing 155
Data source, theme creation from 122
GetFTab request 158
GetFullName request 158
Layer object creation 157
Librarian object 155
Library object 155
Make request 156, 158
Spatial database manager 155, 157
SrcName object 157, 158
VTab object 158
ArcView
Applications 6
RPC server, ArcView as 243
Arrow, adding 194
ASeriesLabel object 168
Assign_Territory 132
AsString request 75
Attribute data
Table documents, programming 131–144
See Also Displaying.
Attribute fields, associating 219
Availability, variables 52
Avenue programming language 131
Accessing files 67, 69, 71
Active objects, checking for 63
AML, distinguished from 38
Applications script creation 81
Calling other programs 66
Characteristics 38
Class hierarchy 245–254
Classes 39
Closing files 69
Comment lines 61
Condition, checking for 62
Conditional structures 57
Controlling program flow 56
Creating or opening files 69
Dictionaries 55
See Displaying.
Documenting with comment lines 61
Elements 51–59
Encapsulation 38
Error message 48
Errors display 72

Index 261

Event programming 61, 63
File name object 68
File name, getting 75
Geocoding, adding event table 211
Geocoding, address matching 211
Holding objects in variables 52
Information display 72
Inheritance 38
Input, getting 73
Interaction among programs 64, 65
Lists 55
Literals 53
Loop structure 58
Name dictionaries 55
New object, creating 40
Nonserial performance of tasks 7
Objects 39
Passing objects among programs 66
Polymorphism 38
Programming environment 45
Programming view documents 107, 109
Reading a file 71
Requests, defined 40
Requests, making 41
Reserved words 255, 256, 257
Run-time errors, correcting 49
Running the script 49
Script compiling 48
Script Editor 44–46
Script testing 47, 49
Script, application creation 95
 See also Scripts
Stacks 55
Starting object 46
Structured methodology development,
 construction, scripts 6
System Editor use 45
System scripts, accessing 47
User dialog 72–75
Warnings display 72
Writing a file 71
See also File based database objects; Integration
 capabilities; Layout documents;
 Themes, programming
Axis 168

B

Bit map object 171
Boolean condition
 Controlling program flow 56
 If statements 57
Box selection, feature selection in themes 129
Break statements 60
Buttons 21

C

Calculate request 152
CanEdit request 147
CanMake request 147
Case study
 Construction 7
 Prototyping 5
 Requirement study 3
 Structured testing 8
Category option 14, 21, 27
Chained requests 43, 54
Chart documents, framing 194
Chart documents, programming
 ChartDisplay class 163, 164
 ChartLegend class 163
 ChartPart class 165
 Creating chart 163
 Data elements, working with 170, 171
 Fields, accessing 172
 Printing charts 173
 Properties, setting 167
 Record, accessing 170
 Record, finding by matching string 173
 Record, identifying 172
 Title class 163
 Type and style, setting 165
 VTab, use 163, 165, 167, 169
 XAxis class 163
 YAxis class 163
Chart properties, setting
 AChartLoc 169
 AGroupLabel object 168
 AGroupNumber object 168
 ASeriesLabel object 168
 Axis 168
 XAxis 168
 YAxis 168

262   Index

Chart property dialog box 164
Chart type and style, setting
   ChartDisplay object 165
   Get requests 166
   GetStatus requests 167
   IsOK requests 167
   SetStyle 165
   SetType 165
ChartDisplay class 163, 164
ChartDisplay object 165
ChartLegend class 163
ChartPart class 165
Choice Message Box 15
Choice request 75
ChoiceAsString request 75
Circle, CRM_Analysis 112
Classes
   Avenue 40
   Avenue class hierarchy 245–254
   GraphicList class 187
   GraphicShape class 187, 189
   GraphicText class 189
   New objects, creating 40
   Obd 202
   Printer class 198
   Requests made to 40
   System class 236, 237
Click events 61
Click property line 23
Client, DDE conversation 238
Clipboard objects 234
Clipboard, accessing
   Add request 234
   Clipboard objects 234
   Empty request 234
   Get request 234
   Graphics 235
   Integration 234, 235
   String objects 235
   Update request 235
Closing documents, project window 84
Closing files 69
Closing tables 139
Cold links, DDE conversation 238
Collections, classes of 55
Comment lines 61
Comments 83, 140
Commit requests 203
Compile pushbutton 48

Compose map, application creation,
   menu items 89
Condition, checking for 62
Conditional structures
   Controlling program flow 56
   If statements 57
Construction
   Case study 7
   Structured methodology development 6
Continue statements 60
Control bar elements, interface customization 12
Control Editor 14, 18–33
Controlling program flow
   Boolean condition 56
   Conditional structures 56
   If statements 57
   Loop structure 56, 58
   True and False objects 56
ConvertRowToRecord request 151
Corporate databases, address matching 222
Creating a layout
   Graphic list 184
   Make request 184
   Retrieval of display 184
Creating charts
   Chart property dialog box 164
   Make request 163
   VTab 163
Creating documents, project window 84
Creating files 69
Creating Table document 137, 138, 139
Creator field 83
Creator, setting table properties 140
CRM_Analysis 110, 112
CRM_Get_Locations 106, 110
CRM_New_Analysis 106, 110, 114
CRM_New_View 106, 107
CRM_Scale 1 106
CRM_Scale 2 106
Cursor line 30
Cursor shape, associating to tool buttons 30
Customize dialog box 13–27, 32, 34, 47, 183
Customized ArcView interface
   Project window programming 80
Customized interface, prototyping 6
Customized pushbutton,
      Layout documents, programming 175

Index   263

D

Data elements, working with
  Fields, accessing 172
  Printing charts 173
  Records, accessing 170
  Records, finding by matching string 173
  Records, identifying 172
Data management 39
Data source
  Make request 122
  SrcName objects 122, 123
  Theme creation from 122
Database access and analysis
  Geographic dimension 222
Databases, accessing
  ArcStorm 155, 157
  File based database objects, creating 145–158
  Records and fields, reading 150, 151
  SQL database, accessing 153
  VTab, accessing database with 150
DBase file class 150
DDE, implementing
  ArcView as DDE client 239
  ArcView as DDE server 240
  Client and server 238
  DDEClient object 239
  Integration 233
  Make request 239
  Start request 240
  Stop request 240
DDEClient object 239
Declaration of variable 52
Default preference values 230
Default project file 204
Defaults, changing 34
Defects, structured testing 8
Delete request 70
Delete, disabling 17
Demonstrations, structured methodology
  development 5
Design alternatives, prototyping 4, 5
Dictionaries 55
Disabled property line 32
Disabling menu item, interface customization 17
Displaying layout 184
Displaying tables
  DocWin object 139
  EditTable request 138

Opening, closing, resizing and moving 139
Displaying themes 119–125
Displaying view document 109
Distributing objects
  Add requests 203
  Commit requests 203
  Get requests 203
  Obd class 202
  Object database 202
  Saving to disk files 202
DocFrame objects 194, 195
Document windows, arranging 84
Documenting with comment lines 61
Documents, application creation tutorial 89
DocWin object 139
Dynamic Data Exchange, integration 233
Dynamic segmentation events,
  Theme creation from data source 122

E

Edit request 197
EditTable request 139
ElementCount 71
Embedding scripts 200
Empty request 234
Encapsulation 38
Encrypting scripts 201
EncryptScripts request 202
EndBatch request 188
Environment variables, accessing
  GetEnvVar requests 236
  Point object 237
  ReturnScreenSizePixels request 237
  SetEnvVar requests 236
Error message 48, 49
Error request 42
Errors
  Avenue programming language,
    testing for run-time errors 48
  Display 72
Event programming 18, 25, 32
  Active objects, checking for 63
  Apply events 61
  Click events 61
  Condition, checking for 62
  GetActive requests 63
  Is requests 62
  IsAtEnd 62

264   Index

Update events 62
Event table, geocoding with Avenue 211
Events
  Address events theme creation 226–232
  Event table selection 227
  Geocoding, adding address events 210
  Geocoding, creating events theme 210
  Unmatched events, processing 210
Examine Variables button 50
Execute request 237, 242
Exists request 70
Export request 143, 149
Exporting records 149
Extent of displayed map, setting 110

F

False value 25, 32
Feature selection
  Box selection 129
  GetMouseLoc request 129
  Line selection 129
  Point selection 128
  Polygon selection 130
  ReturnUserRect request 130
  SelectByPoint request 129
  SelectByRect request 129
Field names 219
Field values, accessing 150
Fields, accessing chart fields with GetFields
request 172
File based database objects
  CanEdit request 147
  CanMake request 147
  Creating 145–158
  DBase file class 150
  DText file class 150
  Export request 149
  Exporting records 149
  HasError request 147
  INFO file class 150
  IsEditable request 147
  Make request 147, 150
  MakeNew request 150
  SetEditable request 147
  VTab 146, 147
File classes, file based database objects 150
File menu option 82
File name, getting 75

FileDialog 75, 138
FileName class object 75
FileWin class 75
Find request 154, 173
Finding documents, project window 84
FindScript request 65
Fire menu item properties 81
For Each loop 58, 59
Framing view document
  GetBounds request 191
  GraphicGroup 194
  LegendFrame objects 193
  Legends, adding 193
  Make request 190, 192, 193
  North arrow, adding 194
  NorthArrow objects 194
  Scale bar, adding 192
  ScaleBarFrame objects 192
  ViewFrame objects 190
Framing view document, Make request 193
Friendliness, interface customization 11
Functionality documentation, prototyping 6

G

Geocode theme, initializing
  GeoName object 228
  MatchSource 228
  Rematch request 228
  Search request 228
Geocoding
  Address events, adding 210
  Avenue, adding event table 211
  Avenue, address matching with 211–215
  Events theme, creating 210
  Linear format 209
  Point format 209
  Polygon format 209
  Unmatched events, processing 210, 211
  Zone information 208, 209
GeoName object 228
Get request 166, 203, 234
GetActive request 63
GetConnections request 154
GetEnvVar request 236
GetFields request 172
GetFTab request 158
GetFullName request 158
GetMouseLoc request 129

Index 265

GetName request 75
GetNumRecords request 151
GetPos request 72
GetRecordFromClick request 172
GetSelection request 171
GetSize request 72
GetStatus request 167
GIS applications 1, 6
Global variables 52
Graphic formats 197
Graphic list, using 115
   AddBatch request 188
   Creating a layout 184
   EndBatch request 188
   GraphicList class 187
   GraphicShape class 187, 189
   GraphicText class 189
   Picture, framing 195
Graphical elements, drawing on existing
   view 115
Graphical User Interfaces
   Help message display 85
   Structured methodology development 1
GraphicGroup 194
GraphicList class 187
Graphics, accessing clipboard 235
GraphicShape class 187, 189
GraphicText class 189
Guide New Layout script, application creation 101
Guide Print script 100
Guide Select States script 96
Guide Update Chart script 98
Guide View Chart script 99
GUIs. *See* Graphical User Interfaces

H

HasError request 147, 242
HasServer request 242
Help dialog box 26
Help line
   Pushbuttons, adding to 25
   Tool buttons, adding 32
Help message display 85
Hiding menu item, interface customization 17
Holding objects in variables 52
Hot links, DDE conversation 238
Human interaction 11, 39

I

Icon Manager dialog box 22, 28
Icons, arranging 84
Identify window 151
If statements, conditional structures 57
Image files, theme creation from data source 122
Import, interface customization 17
Infix, making requests 41
INFO file class 150
Information display 72
Inheritance 38
Input request 75
Input, getting 73
Installing application
   Distributing objects 202
   Modifying scripts 200
   Network installation 205
   Protecting scripts 199–204
   Single user installation 203
Integration
   AMLs, executing 237
   ArcView as DDE client 239
   ArcView as DDE server 240
   ArcView as RPC client 241
   ArcView as RPC server 243
   Clipboard, accessing 234, 235
   DDE, implementing 238, 239
   Dynamic Data Exchange 233
   Environment variables, accessing 236
   Operating system, accessing 236, 237
   Remote Procedure Call 233
   RPC, implementing 240–243
Interaction among programs
   Calling other programs 64
   FindScript request 65
   Master script 64
   Modular design 64
   OwnerObject 66
   Passing objects among programs 66
   Run request 64
   Subroutines 64
Interface customization
   Access keys, assigning 19
   Application creation 90
   Category option 14
   Choice Message Box 15
   Complexity 11–34
   Control bar elements 12

266   Index

Control Editor area 14
Customize dialog box 13, 14, 15
Disabling menu item 17
Friendliness 11
Invisibility 17, 18
Label line 15
Menu 14
Menu bar 12
Menu item, disabling 17
Menu item, hiding 17
Menu options, adding and organizing 14
Names, changing 15
New Item button 15
New Menu button 14
Project window, Customize dialog box 13
Properties list 15
Pushbutton bar 12
Pushbuttons, adding and organizing 21
Pushbuttons, adding help line 25
Pushbuttons, creating 20–25
Pushbuttons, disabling 24
Pushbuttons, hiding 24
Pushbuttons, task execution through 23
Saving customization 34
Separator button 15
Shortcut keys, assigning 19
Shutdown scripts 33
Startup scripts 33
Task execution through menu 16
Tool bar 13
Tool buttons, adding and organizing 27
Tool buttons, adding help line 32
Tool buttons, associating cursor shape to 30
Tool buttons, creating 26–31
Tool buttons, disabling 31
Tool buttons, hiding 31
Tool buttons, task execution sequence
   through 29
Type option 14
Invisibility
   Interface customization 17, 18
   Property line 32
Is requests 62
IsAtEnd 62
IsEditable request 147
IsMachine request 242
IsOK requests 167
Item, DDE conversation 238
Items options, names 16

J
Join request 141
Joining tables
   Join request 141
   Link operation 142
   VTabs 141

K
Keys, dictionaries 55

L
Label dialog box 16
Label line 15
Layer object creation 157
Layout documents, programming
   Avenue scripts 175
   Chart document, framing 194
   Creating a layout 183, 184
   Customized dialog box 183
   Displaying layout 184
   DocFrame objects 194, 195
   Framing view document 190–193
   Graphic list, using 187
   Graphics, adding 187
   Grid, setting 186
   Legends, adding 193
   Maps 175–198
   Name 187
   Page properties, setting 185
   Picture, framing 195, 197
   PictureFrame object 195
   Printer class 198
   Printing layout 198
   Properties, setting 185
   Scale bar, adding 192
   Table, framing 194
LegendFrame objects 193
Legends
   Adding 193
   Creating 124
   Properties, setting 124
Librarian 155
Librarian object 155
Library object 155
Line selection, feature selection in themes 129
Linear geocoding format 209
Link operation, joining tables 142

Index   **267**

Literals
  Creation 54
  String literals 54
Load Script dialog box 45, 46
Load System Script 47
Load Text File pushbutton 45
Load_Att_Table 132
Load_Theme 119
Loadscript 34
Local variables 52
Login request 155
Loop structure
  Break statements 60
  Continue statements 60
  Controlling program flow 56, 58
  For Each loop 59
  For Each loops 58
  Nested 59
  While loop 58, 59

M

Main program 64
Make Default button 34, 81
Make request 108, 122, 147, 150, 156, 158,
    163, 184, 190, 192, 193, 217, 225, 239, 242
MakeNew request 150
Maps
  Setting view 109
  See Also Layout documents.
Margins, SetBounds request 117
Master script 64
MatchCase object 228–230
MatchKey object 228– 230
MatchPref object 228, 230
MatchSource object 217, 225–230
Menu 14
Menu bar, interface customization 12
Menu items
  Application creation 88
  Applications 81
  Direct access 19
  Disabling, interface customization 17
  Hiding, interface customization 17
Menu options, interface customization 14
Menus, task execution 16
Messages, Help message display 85
Microsoft Windows, implementing DDE 233, 238
Modular design, interaction among programs 64

Moving tables 139
MsgBox request 86
MultiInput request 75
Multiple tables, joining 141
Multiple users, network installation 205
Multiple views, setting display extent 110

N

Names
  Applications, naming project 82
  Attribute field names 219
  Avenue reserved words 255–257
  Dictionaries 55
  File name object 68
  Interface customization 15, 16
  Layout documents, programming 187
  Script, creating 44
  Setting table properties 140
  SrcName objects 122
  Variables 52
  View, setting name 109
Naming conventions 6
Nested For Each loop 59
Nested If statement 58
Nested loop structure 59
Network installation 205
New button 27
New Item button 15
New Menu button 14
New objects, creating 40
New view, creating 108
North arrow, adding 194
NorthArrow objects 194

O

Object database
  Distributing objects 202
  North arrows 194
Object oriented languages, Avenue 38
Objects
  Active objects, checking for 63
  AddressStyle object 218
  AGroupLabel object 168
  AGroupNumber object 168
  ASeriesLabel object 168
  Avenue 39
  Bit map object 171
  Clipboard objects 234

268 Index

Collections 55
Controlling program flow, True and False
    objects 56
DDEClient object 239
Dictionaries 55
DocFrame objects 194, 195
See File based database objects.
File name object 68
GeoName object 228
Layer object creation 157
LegendFrame objects 193
Librarian object 155
Library object 155
Lists 55
Literals 54
MatchCase object 228–230
MatchKey object 228–230
MatchPref object 228, 230
MatchSource object 217, 226
MatchSource object creation 225
Name dictionaries 55
NorthArrow objects 194
Passing objects among programs 66
PictureFrame object 195
Point object 237
RPCClient object 241
ScaleBarFrame objects 192
SQLCon object creation 153
SrcName 122
SrcName object 158
SrcName object creation 157
Stacks 55
String objects 54, 243
Tag, setting table properties 140
Variables 52
ViewFrame objects 190
VTab object 137, 155, 158
Open Project menu 82
Open request 109
Opening documents 84
Opening files 69
Opening tables 139
Operating system commands
    Execute request 237
    System class 237
Operating system, accessing
    Environment variables, accessing 236, 237
    Executing AMLs 237
    Issuing operating system commands 237

P

Page properties, setting 185
Paragraph margins, SetBounds request 117
Parameters
    Requests 41
    Setting view parameters 109
Parentheses, Boolean expressions 57
Password request 75
Paste request 126
Picker dialog box 19, 20, 30
Picture, framing
    Edit request 197
    Graphic formats 197
    Picture, framing 196
    PictureFrame object 195
    Refresh properties 196
Point
    CRM_New_Analysis 114
    Geocoding format 209
    SrcName object 122
    Themes, feature selection 128
Point object 237
Polygons
    Assign_Territory 132, 137
    Geocoding format 209
    SrcName object 122
    Themes, feature selection 130
Polymorphism, Avenue 38, 40
Postfix, making requests 41
Prefix, making requests 41
Presentation quality, framing picture 196
Print request 173
Printing charts 173
Printing layout 198
Printing tables
    Export request 143
    Table documents 142, 143
Problem analysis,
    Structured methodology development 3
Product description, structured methodology
        development 3
Project menu option 33
Project menu option, access key assignment 19
Project parameters, applications 83
Project Properties dialog box 34, 83
Project window 13, 26, 34, 45, 159
Project window programming
    Customized ArcView interface 80

Index   269

Tutorial for application creation 88–101
Properties
  Layout properties, setting 185
  Page properties, setting 185
  Setting view properties 109
Properties dialog box 33
Properties list 15–33
Protecting scripts
  Embedding 200
  Encryption 201
  EncryptScripts request 202
  Modifying scripts 200
  ScriptManager dialog box 200
Prototyping
  Case study 5
  Structured methodology development 4, 5
Pushbuttons
  Adding and organizing 21
  Adding help line 25
  Chart documents, executing script 159
  Compile 48
  Creating 20–25
  Disabling 24
  Enabling a customized button 63
  Examine Variables button 50
  Hiding 24
  Interface customization 12
  Run button 48–50
  Step button 48, 50
  Task execution through 23
  Write Text File 45
Put requests 75, 76

Q
Queries 126

R
Reading a file 71
Records
  Exporting records 149
  Finding by matching string 173
Records and field, reading
  AddRecord request 153
  Calculate request 152
  ConvertRowToRecord request 151
  GetNumRecords request 151
  Identify window 151
  ReturnValueNumber request 151

ReturnValueString request 151
SetValue request 153
SetValueNumber request 153
SetValueString request 153
Records, accessing chart
  Bit map object 171
  GetSelection request 171
  VTab 171
Records, identifying chart
  GetRecordFromClick request 172
  Identify tool button 172
Rectangle, CRM_Analysis 112
Refresh properties, framing picture 196
Region.state, SrcName object 122
Rematch request 228
Remote data, accessing 205
Remote Procedure Call. *See* RPC
Rename 44
Report, application creation, menu items 89
Requests
  Add request 234
  Add requests 203
  AddBatch request 188
  AddRecord request 153
  Applications 86
  AsString request 75
  Calculate request 152
  CanEdit request 147
  CanMake request 147
  Chained 43
  Choice request 75
  ChoiceAsString request 75
  Close request 70
  Commit requests 203
  ConvertRowToRecord request 151
  CopyThemes request 126
  CutThemes request 126
  Defined 40
  Delete request 70
  Edit request 197
  EditTable request 139
  Empty request 234
  EncryptScripts request 202
  EndBatch request 188
  Execute request 237, 242
  Exists request 70
  Export request 143, 149
  Find request 154, 173
  FindScript request 65

270   Index

Get request 234
Get requests 166, 203
GetActive requests 63
GetBounds request 191
GetConnections request 154
GetEnvVar requests 236
GetFields request 172
GetFTab request 158
GetFullName request 158
GetMouseLoc request 129
GetName request 75
GetNumRecords request 151
GetPos request 72
GetRecordFromClick request 172
GetSelection request 171
GetSize request 72
GetStatus requests 167
HasError request 242
HasServer request 242
IsEditable request 147
IsMachine request 242
IsOK requests 167
Join request 141
Legends, setting properties 124
Login request 155
Make request 68, 69, 108, 122, 147, 150, 156,
    158, 163, 184, 190, 192, 193, 217, 225, 239,
    242
MakeNew request 150
Making 41
MatchSource request 230
Message display requests 72, 73
MultiInput request 75
Open request 109
Parameters 41
Password request 75
Print request 173
Put requests 75, 76
Read request 71
Rematch request 228
ReturnScreenSizePixels request 237
ReturnUserRect request 130
ReturnValueNumber request 151
ReturnValueString request 151
Run 64
Script execution request 243
Search request 228– 230
SelectByPoint request 129
SetBounds request 117

SetEditable request 147
SetEnvVar requests 236
SetPos request 71
SetStatus request 86
SetValue request 153
SetValueNumber request 153
SetValueString request 153
Show requests 75
ShowMsg request 86
Sort request 142
Start request 240
Stop request 240
Update request 235
Write request 71
WriteMatch request 230
WriteUnMatch request 230
YesNoCancel request 74
Requirement study
    Case study 4
    Structured methodology development 2, 3
Resizing a table 139
Retrieval of display 184
ReturnScreenSizePixels request 237
ReturnUserRect request 130
ReturnValueNumber request 151
ReturnValueString request 151
Route.subclass, SrcName object 122
RPC, implementing
    ArcView as RPC client 241
    ArcView as RPC server 243
    Creating 240–243
    Execute request 242
    HasError request 242
    HasServer request 242
    IsMachine request 242
    Make request 242
    RPCClient object 241
    Script execution request 243
    String objects 243
RPC, integration 233
RPCClient object 241
Run button 48–50
Run requests 64, 66
Run-time errors
    Avenue programming language, testing 48
    Avenue, correcting 49
    Nested and compound If statements 58
Running the script 49

Index 271

S

Save as operation 204
Scale bar, adding 192
ScaleBarFrame objects 192
Script
   Running 49
   Saving 45
   Script Editor 44, 45, 46
   Scripts icon 44, 45
Script compiling 48
Script documentation 6
Script Editor 45, 46, 48
Script execution request 243
Script Manager 47
Script Manager dialog box 16, 17, 23, 29, 34, 200
Scripts
   Application creation 95
   Master script 64
   Modifying 200
   Protecting scripts 200
   See Table documents.
   Testing 47, 49
   See Also Displaying; File based database objects; Interaction among programs; Layout documents; Programming view documents; Themes.
Search request 228–230
Select states, application creation, menu items 89
SelectByPoint request 128, 129
SelectByRect request 129
Selecting theme 217
Separator button 15, 22
Serial number added to script 205
Serial performance of tasks, AML 7
Server, DDE conversation 238
SetBounds request 117
SetEditable request 147
SetEnvVar request 236
SetPos request 71
SetStatus request 86
SetStyle 165
Setting table properties 140
SetType 165
SetValue request 153
SetValueNumber request 153
SetValueString request 153
Shapes, drawing on existing view 116

Shortcut keys, interface customization 19
Show request 75
ShowMsg request 86
Shutdown scripts, interface customization 33
Single user installation
   Default project file 204
   Save as operation 204
   Serial number added to script 205
Sort request 142
Sorting records
   Sort request 142
   Table documents 142
   VTab records 151
Spatial attributes.
   See Address matching
Spatial database manager.
   ArcStorm 155, 157
Spatial elements, manipulating through Table documents 137
Specification document 3
SQL database, accessing
   Find request 154
   GetConnections request 154
   Login request 155
   SQLCon object creation 153
   VTab object 153, 155
SQLCon object creation 153
SrcName object 122, 123, 157, 158
Stacks 55
Start request 240
Starting object, Avenue 46
Startup scripts
   Applications 82
   Interface customization 33
Status bar display 25, 86
Status line messages 86
Step button 48, 50
Stop request 240
Street addresses. *See* Address matching
String literals 54
String objects 243
   ArcView as DDE server 240
   Clipboard, accessing 235
Structured methodology development
   Case study, construction 7
   Case study, requirement study 3
   Case study, structured testing 8
   Construction 6
   GUIs 1

272 Index

Problem analysis, requirement study 3
Product description, requirement study 3
Prototyping 4, 5
Requirement study 2, 3
Specification document 3
Structured testing 8
System testing 8
Unit testing 8
User acceptance testing 8
User participation 4, 5
Structured testing
  Case study 8
  Structured methodology development 8
Style selection 217
Subroutines 64
Superclass 38
Syntax, testing 48
System crash, open files damage 70
System scripts, accessing 47
System testing 8

T

Table documents, programming
  Assign_Territory 132
  Attribute data manipulation 131–144
  Create_Att_Table 132
  Creating Table document 137–139
  Displaying table 138
  FileDialog 138
  Joining tables 141
  Load_Att_Table 132
  Printing tables 142, 143
  Setting table properties 140
  Sorting records 142
  VTab objects 137
Table documents.
  See Table documents, programming
Tables
  Event table selection 227
  Framing 194
Task execution through menu 16
Task execution through tool buttons 29
Task management 39
Testing
  Application creation 102
  Structured methodology development,
    structured testing 8

Text
  Drawing on existing view 116
  SetBounds request 117
Theme, making matchable
  AddressStyle object 218
  Attribute fields, associating 219
  Field names 219
  Make request 217, 225
  MatchSource object 217
  MatchSource object creation 225
  Selecting theme 217
  Style selection 217
Themes
  Address event theme creation 232
  Address events theme creation 226–231
  Avenue, address matching with 211–215
  Geocode theme, initializing 228
  Geocoding, creating events theme 210
  Geocoding, making theme matchable 208
  Setting display extent 110
Themes, programming
  Adding themes 119–125
  Copying 125
  CopyThemes request 125
  Creation of themes from data source 122
  CutThemes request 126
  Data source, theme creation from 122
  Displaying themes 119–125
  Feature selection 127, 129
  Legends 124
  Load_Theme 119
  Queries 126
Title class 163
Tool bar, interface customization 13
Tool buttons
  Adding 27
  Adding help line 32
  Creating 26–31
  CRM_Get_Locations 110
  Cursor shape, associating 30
  Disabling 31
  Hiding 31
  Identify tool button 172
  Organizing 27
  Task execution sequence through 29
Tutorial, application creation 88–101
Type option 14, 21, 27

Index 273

U

Undo button 24
Unit testing 8
UNIX, RPC, transfer between applications 233
Unmatched events, processing 210, 211
Update events 62
Update request 235
Usa coverage, programming table documents 131
User acceptance testing 8
User dialog
  AsString request 75
  Choice request 75
  ChoiceAsString request 75
  Errors display 72
  File name, getting 75
  FileName class object 75
  FileWin class 75
  GetName request 75
  Information display 72–75
  Input request 75
  Input, getting 73
  Message display request 72, 73
  MultiInput request 75
  Password request 75
  Put request 75, 76
  Show request 75
  Warnings display 72, 73
  YesNoCancel request 74
User participation 4, 5

V

Variables 40
  Availability 52
  Avenue programming language 52
  Avenue reserved words, names 255–257
  Declaration 49, 52
  Global 52
  Local 52
  Names 52
View chart, menu items 89
View documents 46
  Programming 106
View documents, programming
  Application overview 105
  CRM_Analysis 110
  CRM_Get_Locations 106, 110
  CRM_New_Analysis 106, 110, 114
  CRM_New_View 106, 107

  CRM_Scale 1 106
  CRM_Scale 2 106
  Displaying view document 109
  Extent of displayed map, setting 110
  Graphical elements 110–117
  Graphical elements, drawing 115
  Graphics list 115
  Make request 108
  Map units, setting 109
  Name, setting 109
  New view, creating 108
  Open request 109
  Parameters, setting view 109
  Properties, setting view 109
  SetBounds request 117
  Shapes, drawing on existing view 116
  Text, drawing on existing view 116
ViewFrame objects 190
VTab
  Chart documents, programming 163
  Creation 146
  File based database objects 147
  Joining tables 141
  Record, finding by matching string 173
  Records, accessing chart 171
VTab object 137, 153, 155, 158

W

Warnings display 72, 73
While loop 58, 59
Window, interface customization 17
Write Text File pushbutton 45
WriteMatch request 230
WriteUnMatch request 230
Writing a file 71

X

XAxis 163, 168

Y

YAxis 163, 168
YesNoCancel request 74

274   Index

Z

Zone field objects 227
Zone information, geocoding 208
Zoomin tool 30

More
OnWord Press Titles

Pro/ENGINEER and Pro/JR. Books

INSIDE Pro/ENGINEER
Book $49.95 Includes Disk

Pro/ENGINEER Quick Reference, 2d ed.
Book $24.95

Pro/ENGINEER Exercise Book
Book $39.95 Includes Disk

Thinking Pro/ENGINEER
Book $49.95

INSIDE Pro/JR.
Book $49.95

Interleaf Books

INSIDE Interleaf
Book $49.95 Includes Disk

Adventurer's Guide to Interleaf Lisp
Book $49.95 Includes Disk

The Interleaf Exercise Book
Book $39.95 Includes Disk

The Interleaf Quick Reference
Book $24.95

Interleaf Tips and Tricks
Book $49.95 Includes Disk

MicroStation Books

INSIDE MicroStation 5X, 3d ed.
Book $34.95 Includes Disk

MicroStation Reference Guide 5.X
Book $18.95

MicroStation Exercise Book 5.X
Book $34.95
Optional Instructor's Guide $14.95

MicroStation 5.X Delta Book
Book $19.95

MicroStation for AutoCAD Users , 2d ed.
Book $34.95

Adventures in MicroStation 3D
Book 49.95 Includes Disk

MicroStation Productivity Book
Book $39.95
Optional Disk $49.95

MicroStation Bible
Book $49.95
Optional Disks $49.95

Build Cell
Software $69.95

101 MDL Commands
Book $49.95
Optional Executable Disk $101.00
Optional Source Disks (6) $259.95

101 User Commands
Book $49.95
Optional Disk $101.00

Bill Steinbock's Pocket MDL Programmer's Guide
Book $24.95

MicroStation for AutoCAD Users Tablet Menu
Tablet Menu $99.95

Managing and Networking MicroStation
Book $29.95
Optional Disk $29.95

The MicroStation Database Book
Book $29.95
Optional Disk $29.95

The MicroStation Rendering Book
Book $34.95 Includes Disk

INSIDE I/RAS B
Book $24.95 Includes Disk

The CLIX Workstation User's Guide
Book $34.95 Includes Disk

SunSoft Solaris Series

The SunSoft Solaris 2.* User's Guide
Book $29.95 Includes Disk

SunSoft Solaris 2.* for Managers and Administrators
Book $34.95
Optional Disk $29.95

The SunSoft Solaris 2.* Quick Reference
Book $18.95

Five Steps to SunSoft Solaris 2.*
Book $24.95 Includes Disk

One Minute SunSoft Solaris Manager
Book $14.95

SunSoft Solaris 2.* for Windows Users
Book $24.95

Windows NT

Windows NT for the Technical Professional
Book $39.95

The Hewlett Packard HP-UX Series

The HP-UX User's Guide
Book $29.95 Includes Disk

The HP-UX Quick Reference
Book $18.95

Five Steps to HP-UX
Book $24.95 Includes Disk

One Minute HP-UX Manager
Book $14.95

CAD Management

One Minute CAD Manager
Book $14.95

Manager's Guide to Computer-Aided Engineering
Book $49.95

Other CAD

CAD and the Practice of Architecture: ASG Solutions
Book $39.95 Includes Disk

INSIDE CADVANCE
Book $34.95 Includes Disk

Using Drafix Windows CAD
Book $34.95 Includes Disk

Fallingwater in 3D Studio: A Case Study and Tutorial
Book $39.95 Includes Disk

Geographic Information Systems/ESRI

The GIS Book, 3d ed.
Book $34.95

INSIDE ARC/INFO
Book $74.95 Includes CD

ARC/INFO Quick Reference
Book $24.95

ArcView Developer's Guide
Book $49.95

INSIDE ArcView
Book $39.95 Includes CD

DTP/CAD Clip Art

1001 DTP/CAD Symbols Clip Art
Library: Architectural
Book $29.95

DISK FORMATS:
MicroStation
DGN Disk $175.00
Book/Disk $195.00

AutoCAD
DWG Disk $175.00
Book/Disk $195.00

CAD/DTP
DXF Disk $195.00
Book/Disk $225.00

Networking/LANtastic

Fantastic LANtastic
Book $29.95 Includes Disk

The LANtastic Quick Reference
Book $14.95

One Minute Network Manager
Book $14.95

OnWord Press Distribution

End Users/User Groups/Corporate Sales

OnWord Press books are available worldwide to end users, user groups, and corporate accounts from your local bookseller or computer/software dealer, or from HMP Direct: call 1-800-223-6397 or 505-473-5454; fax 505-471-4424; write to High Mountain Press Direct, 2530 Camino Entrada, Santa Fe, NM 87505-8435, or e-mail to ORDERS@ BOOKSTORE. HMP.COM.

Wholesale, Including Overseas Distribution

High Mountain Press distributes OnWord Press books internationally. For terms call 1-800-4-ONWORD or 505-473-5454; fax to 505-471-4424; e-mail to ORDERS@ IPG.HMP.COM; or write to High Mountain Press/IPG, 2530 Camino Entrada, Santa Fe, NM 87505-8435, USA. Outside North America, call 505-471-4243.

Comments and Corrections

Your comments can help us make better products. If you find an error in our products, or have any other comments, positive or negative, we'd like to know! Please write to us at the address below or contact our e-mail address: READERS@HMP.COM.

OnWord Press
2530 Camino Entrada, Santa Fe, NM 87505-8435 USA